PRACTICAL FARMING

PASTURE MANAGEMENT

Rick Bickford

INKATA PRESS

INKATA PRESS

A DIVISION OF BUTTERWORTH-HEINEMANN

AUSTRALIA

BUTTERWORTH-HEINEMANN
North Tower 1–5 Railway Street
Chatswood NSW 2067

BUTTERWORTH-HEINEMANN
18 Salmon Street
Port Melbourne 3207

UNITED KINGDOM
BUTTERWORTH-HEINEMANN LTD
Oxford

USA
BUTTERWORTH-HEINEMANN
Stoneham

National Library of Australia Cataloguing-in-Publication entry

Bickford, Richard N. S. (Richard Nevill Stoddart).
Pasture management.

Includes index.
ISBN 0 7506 8913 7.

1. Pastures - Australia. I. Title. (Series: Practical farming).

633.2020994

©1995 Adelaide Institute of TAFE

Published by Reed International Books Australia. Under the Copyright Act 1968 (Cth), no part of this publication may be reproduced by any process, electronic or otherwise, without the specific written permission of the copyright owner.

Enquiries should be addressed to the publishers.

Typeset by Ian MacArthur, Hornsby Heights, NSW.

Printed in Australia by Ligare Pty Ltd, Riverwood, NSW

Contents

1	**Preliminary considerations**	7
	What to consider when sowing or renovating a pasture	10
2	**Pasture grasses**	20
3	**Pasture legumes**	45
4	**Sowing methods**	73
	Germination requirements	73
	Seedbed preparation	74
	Sowing methods	78
5	**Invertebrate pest control in pastures**	85
	Pesticides	85
	Other control methods	87
	Invertebrate pests	88
6	**Weeds in pastures**	107
	Weed control methods and techniques	115
	Herbicides	124
	Application of herbicides	128
	Precautions and safety equipment when using herbicides	131
7	**Fertiliser requirements and grazing management**	134
	Plant nutrients	134
	Soil tests	139

	Fertilisers	145
	Grazing management	146
8	*Fodder crops and conservation*	150
	Fodder crops for summer and autumn feed	151
	Fodder crops for winter feed	155
	Fodder conservation	157
	Conserving fodder as hay	164
	Conserving fodder as silage	170
	List of species	180
	Index	181

1

Preliminary considerations

Pasture management is an extremely complex problem involving a wide range of factors including climatic conditions, soil types, varietal characteristics, topography and available equipment. No matter what the farm enterprises are, pasture management is an ongoing program requiring a wide range of knowledge and skills.

Since the arrival of the First Fleet in 1788, the cargo of which included seeds of pasture grasses and clovers, Australians have been striving to improve and manage their pastures better to make the best use of their most valuable asset, the land. Improved pastures have resulted in increased productivity and returns from both livestock and cropping enterprises, and at the same time been responsible for the maintenance or improvement of soil structure and fertility levels.

The improved pasture productivity has been achieved in a number of different ways including the introduction, selection and breeding of higher yielding varieties, and by developing better farming machinery and cultural practices.

Throughout Australia there is considerable variation in climate and topography which has an impact on the types of pastures produced and the systems used to produce and manage these pastures (Figure 1.1). In broad terms the difference in climate will affect the growth habit of the pasture species and the seasons of the year when they are most productive.

The **warm temperate** zone of southern Australia experiences a mediterranean type climate with moist mild winters and warm dry summers. These areas include the central and south-western slopes of New South Wales, the western parts of Victoria and the less arid parts of Western Australia and South Australia. The winter dominant rainfall is suited to annual pasture species but is insufficient to support most perennial pasture species. Typical growth patterns for pastures commence with rapid growth

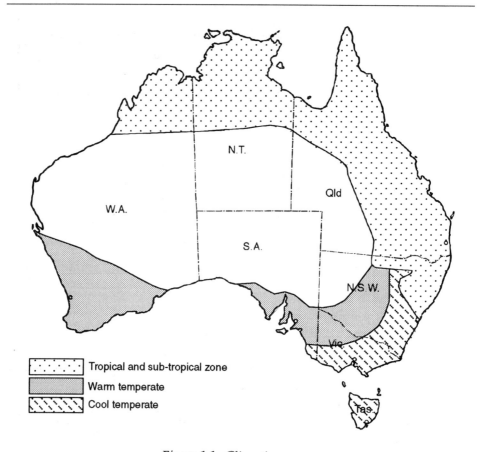

Figure 1.1 Climatic zones

after the opening rains in autumn, followed by slower growth in the cool winter months. The growing season terminates with a spring flush before the annual species set seed and die (Figure 1.2).

Cool temperate climates occur in south-eastern Australia and are typified by a high rainfall pattern and higher altitudes. Rainfall is not normally limiting, but in the summer months high evaporation can result in moisture stress for pastures. Growth in winter months is restricted by low temperatures and waterlogging. Peak pasture growth occurs during spring and early summer, but drops off rapidly as winter temperatures drop. Cool temperate perennial pasture species are most suited to this climatic zone (Figure 1.3).

Tropical and **sub-tropical** climatic zones are those areas with a predominance of summer rainfall. Pasture species in these areas are mostly perennials that grow quickly with summer rainfall, have no appreciable growth in autumn and remain dormant during the dry winter and spring. These pastures have a rapid drop in protein levels and digestibility once growth ceases and livestock are unable to maintain liveweight during these periods of the year (Figure 1.4).

Preliminary considerations

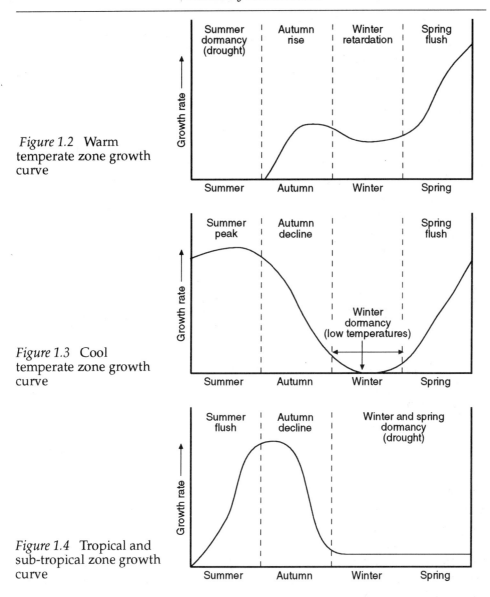

Figure 1.2 Warm temperate zone growth curve

Figure 1.3 Cool temperate zone growth curve

Figure 1.4 Tropical and sub-tropical zone growth curve

It is in the cool and warm temperate zones that the greatest advances in pasture production have been achieved.

In the warm temperate zone, where the rainfall is between 300 and 500 mm (12-20") per annum, the system of Ley farming has been developed. Ley farming involves alternating pasture leys comprising self-regenerating annual legumes with cereal crops such as wheat, barley and oats. The legumes not only increase pasture productivity, they also "fix" atmospheric nitrogen, converting it into nitrogenous compounds in the soil, increasing fertility which in turn boosts the yields of subsequent cereal crops. The protein rich forage material produced by the legumes also enables higher stocking rates of sheep and cattle to be carried on the pastures, and in addition to these benefits the legumes also help to improve soil structure and reduce the incidence of cereal root disease.

This Ley farming system has proved so successful that it is now being widely used in many overseas countries with similar temperate climatic conditions to those of southern Australia, and the exporting of Ley farming technology, farm machinery and pasture seed has developed into a major industry in Australia.

In the cool temperate zone where the average annual rainfall is greater than 500 mm (20"), pastures have been improved by the introduction of perennial grasses that are undersown with either annual or perennial legumes. These pastures are intensively grazed, usually with sheep for meat and wool production and by cattle for the production of beef and dairy products. An increasing range of high value crops such as vegetables, flowers, fruit, oil seeds and pasture seeds are also grown in this zone.

What to consider when sowing or renovating a pasture

There are numerous factors to consider when deciding to sow or renovate, the first an understanding of how various pasture plants, and weeds, grow. Plants can have either annual, biennial or perennial lives.

Annual plants grow only from seed. They germinate vigorously, grow to maturity, flower, set seed and die all within 12 months.

Some early maturing annuals take as little as 3-4 months to complete their life cycle, while some late maturing annuals may take the full 12 months. Those early maturing species are adapted to arid regions where their short life cycle enables them to regenerate reliably. The same plants can grow in higher rainfall areas, but are not as productive as later maturing species. The later maturing species, however, would not be able to survive in the arid regions. They would germinate satisfactorily and start to grow but would run out of time and not be able to flower and set seed because the growing season is too short. These late maturing annuals would therefore not regenerate in areas where the rainfall is too low.

It is therefore essential when selecting annual pasture species that you choose ones that are going to regenerate reliably in the rainfall area that you have in mind, and make full use of the growing season.

Biennial plants take between one and two years to complete their life cycles. They grow from seed and make vegetative growth during the first year, reach maturity, flower, set seed and die during the second.

Perennial plants live for more than two years. Some such as kikuyu grass readily grow from vegetative parts as well as from seed. They do not possess the seedling vigour of annuals and biennials and take longer to establish. Once they reach maturity they generally continue to flower annually until they die, but they do not set seed as prolifically as annuals or biennials.

Perennial pasture species are generally restricted to those districts where the annual rainfall is 400 mm (16") or more, though there are some notable exceptions such as lucerne and perennial veldt grass.

Recognition and understanding of these groups of plants will enable you to make a reasonable assessment of the situation and provide a good guide to the pasture varieties that are suited to the particular site and to the past management of the site.

No two sites are the same when it comes to sowing or renovating pastures and to complicate the situation even more, site conditions are constantly changing. Pasture management is a continuous part of the management of any property whether the enterprise be mainly grazing, cropping, hay or small seed production.

Pastures usually require renovating or resowing when they are producing poorly, infested with weeds or following a crop rotation. Poor production and weed infestation may be the result of many different reasons including destruction by bushfires or overgrazing, drought, insects or disease, low soil fertility or just poor management.

Successful pasture renovation and sowing programs depend on forward planning and careful consideration of many different factors.

Factors to consider are listed as follows in an attempt to illustrate the complex nature of the job. Remember each pasture will be different and the situation is constantly changing and the relative importance of the factors listed will vary accordingly.

Site

Measure the **area** of land to be treated to determine the quantities of seed and fertiliser required. The area may also influence the size and types of equipment to use, and perhaps whether it would be more economical to use contractors for the job rather than purchasing equipment. The area must be known to determine costs. It can be calculated from land title maps, other scale maps of the property, or may be determined from previous records.

The **topography** of the land is an important consideration. It may influence where and when machinery can be safely used, and what is appropriate. Areas with different topography may require different management and different pasture species and may need to be fenced accordingly.

Aspect should be also considered. The southern slopes of hills usually have a longer growing period than northern slopes and may support later maturing pasture varieties.

Inspection of the site will reveal if the site is **arable**. This is most important as it determines the methods that can be used to either sow or renovate the pasture and the most suitable equipment. It is only possible to prepare a good seed bed and maximise the chance of a good pasture establishment on arable land. Less favourable seed beds may restrict the pasture species that can be sown on the land, and will certainly increase the loss factor, necessitating a higher seeding rate to compensate for the losses.

Existing plant cover will provide the best indications of the pasture plants that are thriving or surviving and that are obviously suited to the area. It will also show up weed populations and the proportions of pasture species, weeds and bare, non-productive ground.

It is necessary to take seasonal conditions into account when making paddock inspections, and to look carefully to gauge seed reserves that may or may not influence the regeneration of annual pasture species and weeds.

Inspection of adjoining paddocks and properties will also often assist in gauging the standard and potential of the pasture and help determine future management decisions.

Climatic conditions

Rainfall is of vital importance. The annual rainfall and its seasonal distribution largely determines the range of perennial pasture varieties that are likely to persist and the annual pasture varieties that will make full use of the growing season, yet regenerate reliably.

Most pasture seeds germinate best at soil **temperatures** between 15°C and 25°C. It is therefore wise to sow seed in the autumn or spring when these conditions prevail. Avoid sowing in winter when temperatures are too low for germination, or in summer when sowing temperatures are likely to dry soils out quickly killing germinating seed and young plants unless irrigation is available.

Seasonal production and dormancy of many pasture plants also varies greatly with temperature, and mixtures of different pasture varieties should be chosen to spread production as much as possible.

In areas prone to wind **erosion** cover crops may have to be sown first to protect germinating pasture plants and seed bed preparation may have to be modified to leave the soil surface covered with trash to minimise it.

Sowing should occur early in the growing season to avoid frost damage to young pastures. Usually, in frost prone areas, the selected pasture species are tolerant to frosts but growth and germination are severely restricted due to low soil temperatures. In some instances, severe frosts will cause the soil crust of cultivated paddocks to lift resulting in damage to the roots of young pasture plants. This is commonly known as "frost heave".

Renovation programs are frequently required following periods of drought to restore positive productivity and to prevent weeds infesting the bare areas left by the drought.

Soil

Soil type is important as most pasture varieties grow best on particular types of soil, and will not produce well or persist on others.

Soils with a poor **structure** may require renovation before establishment of a new pasture. Renovation of soil structure may be achieved mechanically through subsoil cultivation with a deep ripper or implements such as the "agro plow", or chemically through the incorporation of soil conditioners such as gypsum, lime or organic matter. It is wise to seek professional advice before applying chemical soil conditioners to the soil. The structure of a soil is usually an indication of its previous management. To achieve a long term improvement in the soil's structure it may be necessary to change management practices that have been the cause of poor soil structure. This may include changes to cropping, cultivation and grazing practices.

The degree of alkalinity or **acidity**, the pH, of a soil will determine which pasture plants, particularly legumes, will thrive in the area. It also affects the availability of plant nutrients which in turn affects pasture production, so steps should be taken to correct extremes of acidity and alkalinity.

The soil's **fertility** will have a big bearing on pasture productivity. A soil's fertility level is determined by a whole range of complex factors. As mentioned previously, pH is one of these factors as well as its organic matter level, type of parent material, drainage, aeration, soil microbial activity, previous cropping and pasture history and fertiliser usage. In addition to all of these factors it is uesful to have a soil test carried out to assist in making decisions on the management of soil fertility for pastures. This will usually require some form of professional advice.

Depth of topsoil is important for healthy plant growth, particularly for deep rooted perennial varieties such as lucerne. It is generally recognised that lucerne requires well drained soils at least one metre deep. Shallow soils, where the clay is often close to the surface, are often poorly drained. The depth of topsoil should be gauged before selecting pasture varieties to sow.

While most pasture varieties require well drained soils for healthy growth others will grow in poorly drained areas and even withstand short periods of waterlogging.

Saline areas need to be sown separately to salt tolerant pasture varieties to prevent further degradation of the salt affected area.

Varietal characteristics

Nutritive value, digestability and palatability all vary between different pasture varieties, different parts of the individual plants and different stages of growth. Pasture mixtures should be blended to aim at producing a supply of fresh green high protein fodder that will remain palatable and nutritious for as long as possible. A good pasture blend is difficult to achieve because usually the more dominant species in the pasture will eventually out compete the less dominant species. Grazing management, fertiliser management and sod seeding replacement species are methods of keeping the pasture blend as required.

Productivity is naturally important and pasture species should be selected not only to produce high yields but to spread seasonal production.

Vigour is an important characteristic to consider. Pasture species should possess good seedling vigour to allow them to establish quickly, compete in the sward and recover quickly from grazing or cutting. Lack of seedling vigour restricts establishment of some species, particularly perennials, and makes them unsuitable for sowing in any situation other than a prepared seed bed.

The **habit of growth** of the different species should also be considered. Species with upright growth habits generally have to be sown at higher rates to obtain an effective ground cover while species with prostrate or bushy growth habits require lower sowing rates. Care must be taken to allow annual pasture species to set sufficient seed to regenerate the following year by not grazing or cutting them too low when they are flowering and setting seed.

Longevity of different perennial species varies greatly and depends to a large degree on good management of species that are suited to a particular site. There are many perennial pastures that are at least ten years old and still producing well, while there are others that require resowing after only a few years.

Resistance or tolerance to disease and insects must also be considered, particularly when choosing legumes. Avoid sowing susceptible species wherever possible.

Toxicity can be a problem with some pasture species at certain stages of growth and under certain conditions. When choosing different species find out if they are likely to cause any problems to animal health.

The ability of different species to withstand extreme conditions such as prolonged high temperatures, drought, frost and waterlogging, set grazing and close grazing must also be considered.

None of these different factors should be overlooked when making a choice of varieties. Make sure to ask your local seed merchant, experienced neighbours, or Department of Agriculture officers if you have doubts about the suitability of any pasture species for a particular use in a particular area.

Timing

Timing is important. Seed must be sown when the soil is damp and likely to remain damp until the seed has germinated and established, and when the soil temperature is favourable for germination. Autumn is usually the most suitable time for sowing in southern Australia as there is a good chance of sufficient rainfall occurring to achieve germination and establishment and soil temperatures are still high enough. Spring sowing can be achieved in areas where summer rainfall is more reliable or where supplementary irrigation is possible. In some coastal areas of southern Australia, spring sowing of summer growing species is common.

On occasions when sufficient rainfall does not fall at the critical sowing time it may be better to postpone sowing the pasture until the following year. Pasture seed is expensive and with the other costs associated with pasture establishment, a failure can be costly.

Competition

Competition from weeds can adversely affect newly sown pastures. Weed control should begin in the year prior to establishment with a series of weed reduction strategies including heavy grazing, application of selective and non-selective herbicides, spray grazing, spray topping, mowing and cultivation. Where new species are being established in existing perennial pasture swards, grazing management and selective herbicide applications are necessary after sowing.

Protection

Newly sown pastures need to be protected from wind erosion, water erosion and insect damage. On light sandy soils where wind erosion is likely to destroy seeding pasture plants, cover crops should be considered. Leaving the ground rough, working it on the contour, and sod seeding rather than cultivating are measures to consider to avoid erosion. Constant close inspection and quick action is the only way to prevent insects and mites causing unnecessary damage.

With a number of invertebrate pasture pests it is necessary to have a management program in place 12 months before sowing the pasture. This is necessary to reduce the pest population to such low levels that it will cause minimal damage when newly established pastures are most susceptible.

Equipment

The available equipment often determines the success of a pasture renovation or sowing program.

If the equipment is available at the right time to kill the weeds and prepare a fine even seed bed and to sow the seed evenly spaced at a constant depth that is not too deep, then the seed will have an excellent chance of germinating and developing into a dense even stand.

If, however, cultivation equipment is not available, and the area is to be sod seeded then the placement of the seed will be more erratic, competition from established plants greater and the mortality rate higher. Seeding rates need to be increased, grasses by up to 20 per cent to compensate for this increased mortality. Varieties with poor seedling vigour should be avoided.

Sod seeding techniques and machinery have improved substantially in recent years and the chances of successfully establishing pastures through these methods have increased. Sod seeding and direct drilling of pastures also have the added advantages of being less damaging to the soil structure, cheaper on fuel and machinery wear and tear, and land is out of production for less time leading up to the point of sowing. This method does require the use of more herbicides.

Broadcasting as a method of establishment is the least likely to succeed and is only recommended on land that is not suited for cultivation and direct seeding. Preparation for aerial or ground seeding of pastures by broadcasting is crucial and should commence 12 months before sowing.

Land management

Many lessons can be learnt from the previous management of the area. Pasture species that have survived or done well should be used again while those that have failed are either not suited to the area or the previous land management. Check the characteristics of these varieties, but do not use them again unless altering the land management techniques.

Consider the grazing needs of the pasture before starting the renovation or resowing program. Select pasture species that are going to suit the livestock enterprises and any other enterprises on the property. Cattle and horses do not graze as closely as sheep. Horses are very selective grazers and tend to cause extensive damage to pastures pulling plants up and cutting them out with their hooves.

Horses also require more fibre in their diets than cattle and sheep and the sowing ratio of grasses to legumes should be 80:20, compared with a ratio of 60:40 grasses to legumes for sheep and cattle.

Effort should also be made to choose strong rooted perennial species that are going to withstand the horses' hooves and survive, even at the expense of productivity, where only horses are concerned.

Costs

The cost of pasture establishment is high and should be regarded as a medium term investment for improving farm productivity. The cost of establishing a highly productive pasture can be in the order of $180 per hectare. Seed, as a component of this cost will vary substantially with species and varieties but it is not uncommon to incur seed costs in the order of $30–40 per hectare for common pasture species at recommended sowing rates. Other pasture establishment costs include herbicides, insecticides, machinery operation costs and fertilisers. It may also be necessary to budget for the cost of purchasing additional livestock to cope with the improved carrying capacity of the pasture.

It is not uncommon for pasture improvement projects to take 4–5 years to break even. This will be subject to a number of factors including livestock and commodity prices, but it is possible that longer periods of time may be needed to break even. In 1992 when the wool price indicator was around 450 cents per kilogram clean, it was not possible to break even on a pasture improvement project. Before commencing a pasture improvement operation it is necessary to budget for the project to determine likely financial benefits.

Pasture improvement also has many non-financial benefits that must also be considered. Improving pastures will reduce land degradation, reduce weed infestation, increase the long term productivity of the farm and increase the quality and quantity of livestock production. The manager may also have improved flexibility in managing livestock and coping with drought. Property value may increase substantially with improved pastures.

Seed quality

Seed quality is the key to successful pasture sowing.

When purchasing seed there are three main factors to consider:

- varietal purity
- germination
- freedom from weeds.

Varietal purity

Varietal purity is guaranteed by the purchase of Certified Seed.

Because the seed of many pasture varieties looks almost identical, the Departments of Agriculture in the various Australian states, and in many overseas countries, run certification schemes to produce pasture seed.

The scheme involves the relevant Department providing mother seed of a known variety to a registered grower, supervising the sowing, growing and harvesting of the seed crop, then the cleaning, packaging, and sampling and testing of the seed, and finally issuing a Certificate of Analysis for the particular line of seed and corresponding certification tags sealed onto the sacks.

Seed displaying the certification label is guaranteed "true-to-name", free of noxious weeds and of high purity and germination. It may be a little more expensive because of the additional production costs but the extra cost is a small price to pay for the guarantee of quality seed compared to the risks of buying inferior seed.

Certification labels (tags) clearly display varietal name, and physical standards such as purity and germination, inert matter, and if present, undesirable weeds such as dock.

The Certificate of Analysis details all the above information plus other seeds, so it is best and safest to buy on analysis.

Analytical certificates should be available from seed merchants. If they are not it will be necessary to obtain them from the section of the Department of Agriculture which administers certified seed production.

Although most seed offered for sale in Australia, both certified and uncertified, has been tested, prospective buyers should always insist on seeing a copy of the relevant analytical report. If there is any doubt as to the authenticity of the report, it is wise to send a sample of the seed to a seed testing laboratory for analysis.

The object of the purity analysis is:
- to determine the composition by weight of the sample under test and by inference the composition of the seed lot
- to identify the different seed species and inert particles constituting the sample.

Pure Seed; establishes the principal species in the sample. (Correct naming is the responsibility of the sender.)

Other Seed; includes any other seeds which are either cultivated as crops or recognised by law or regulation as weed seeds.

Prohibited Seed; State laws prohibit the sale of seed lots containing prohibited seed.

Declared Seed; seeds proclaimed by law or regulation as declared seeds. Sale is not prohibited but the vendor is bound to declare the degree of contamination per kilogram.

Inert Matter; includes all other material not seed such as plant material, soil, stones, small pieces of broken seed, fungal bodies such as ergots or sclerotinia and nematode galls.

Germination

The germination test determines the maximum planting potential of the seed, given favourable conditions of moisture, temperature and a well prepared soil. (Laboratory conditions are favourable.)

Figures shown on the analysis report are:

Normal Seedlings. The percentage of seed producing seedlings capable of continued development into normal healthy plants in favourable conditions.

Hard Seed. Seeds of legumes (clovers, medics, lucernes, peas etc.) which remain hard at the end of the prescribed test period, their impermeable seed coats prevent absorption of water. They are viable but may not germinate immediately after planting.

Fresh Ungerminated Seed. These seeds appear fresh and firm at the end of the test, they have absorbed water but failed to germinate. Some may germinate under favourable conditions. They must never be included in the total germination percentage.

Abnormal Seedlings from old, damaged or poorly developed seed. They do not show the capacity for continued development into normal plants in the field. They should be ignored in anticipating the potential planting value of the seed.

Dead Seed. Seed which has broken down and decayed during the test period.

Potential planting value

Key figures to be used in anticipating the potential field value of the seed lot are:
- Percentage pure seed
- Percentage normal seedlings
- Weed seed content.

The range of variables that can have an impact on the success of a pasture renovation or establishment program indicate that considerable planning is necessary. Failure to consider any one of these variables could result in the pasture improvement program not achieving the desired level of production. If all of these factors are given due consideration, and the planning process is thorough, then only the weather may prevent success.

2

Pasture grasses

Before the introduction of improved pasture species, many of the pastures in Australia consisted of volunteer grasses, herbs and legumes. Many of these had little nutritive value; they were unpalatable to livestock and produced only small yields of herbage. Some also harboured cereal diseases and produced seeds that were capable of contaminating wool and damaging hides.

In the warm temperate zones and drier areas of Australia these problems can be largely overcome by planting and growing vigorous stands of legumes, but in the higher rainfall areas, and where irrigation is available, the unwanted volunteer species can generally be replaced by perennial grasses.

While most perennial grass species are restricted to those areas where the annual rainfall is at least 400 mm (16"), perennial veldt grass is an exception. It will grow in areas where the rainfall is as low as 350 mm (14") per annum (Figure 2.1).

The inclusion of perennial grasses in improved pastures serves two main purposes. Firstly, once they are established, they will compete strongly with the invading annual weeds, and secondly, because of their long growing periods, they are able to increase the production of pasture and spread this production over a longer period than would be possible with annual varieties.

In mixed pastures, perennial grass species should generally be used to make up two thirds of the mixture, but this proportion may need to be altered to suit special requirements such as grazing horses, seed production or hay requirements.

New pasture species and cultivars are constantly being researched in Australia and throughout the world. As a result of this it is possible that the

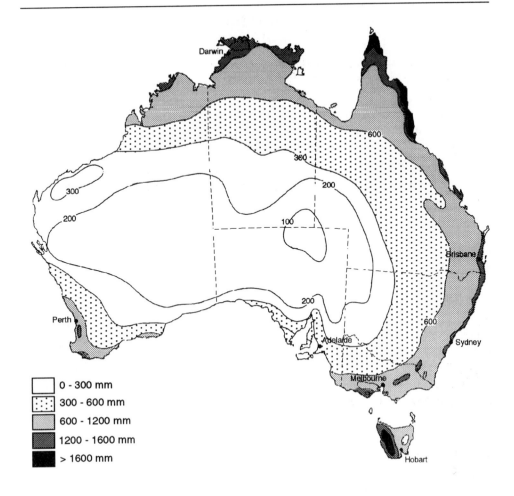

Figure 2.1 Annual rainfall

pasture cultivars mentioned below are not the only cultivars available in Australia and that many new cultivars will eventually be available. Some pasture species are also not mentioned here. These include the vast range of native species that are being selected for commercial pasture mixes. It is likely that within a few years grasses such as wallaby grass (*Danthonia* spp.), kangaroo grass (*Themeda australis*) and weeping grass (*Microleana stipoides*) will be included in pasture mixes.

Other pasture species not included in the following text are the tropical and the sub-tropical species. Pasture improvement in these climatic zones is not as widespread as it is in southern part of Australia, but in the past two decades there has been an increasing level of research into all aspects of pasture improvement in these climates. There are a number of widely accepted tropical and sub-tropical species which include both grasses and legumes.

Cocksfoot
Dactylis glomerata

Cocksfoot cultivars rank third in importance behind ryegrass and phalaris as perennial grasses sown in high rainfall and irrigated pastures in southern Australia.

Native to Europe, northern Africa and temperate Asia cocksfoot has made a significant contribution to Australian pastures through its ability to extend the main period of green feed supply into the summer months.

Cocksfoot is a stout tussocky deep rooted perennial grass with light green to blue-green foliage. The leaves which are folded in the bud (Figure 2.2) have a smooth upper surface, but because of a prominent midrib they feel rough to touch underneath.

To grow well, cocksfoot needs fertile well drained soils, doing best on rich loams. It is very palatable but cocksfoot is not as nutritious as ryegrass and if allowed to grow tall it quickly loses its feed value. Grazing pressure should be regulated to prevent stands from becoming too tall and rank, but at the same time it should not be grazed to ground level for long periods as it may not persist. This is more likely to occur when sheep are grazing.

Cocksfoot has one big advantage over the phalaris and ryegrass cultivars. It does not at any stage of growth cause toxicity problems. It is perfectly safe to graze wherever sufficient herbage is available.

Cultivars of cocksfoot can be divided into two distinct groups that look very similar, but have different growth characteristics. They are either northern European or Mediterranean types.

Northern European types
Aberystwyth S26 (known as S26) cocksfoot
Grasslands Apanui cocksfoot
Porto Cocksfoot

These cultivars are more winter dormant than the Mediterranean types and are used only in irrigated pastures and those districts where the annual rainfall exceeds 650 mm (26") and summer temperatures are mild.

Under irrigated conditions they produce well throughout the spring, summer and autumn, but their winter production is limited.

They can be sown in either autumn or spring and best results are obtained when they are sown 1 cm deep into a well prepared seedbed, at rates of 2–4 kg/ha.

S26 was developed by the Aberystwyth Plant Breeding Station in Wales while Grasslands Apanui was developed by the Grasslands Division of the New Zealand Department of Scientific and Industrial Research.

There is little to choose between S26 and Apanui, although Victorian trials favour Apanui.

The Tasmanian Department of Agriculture developed Porto cocksfoot from seed originally introduced from Portugal by the CSIRO.

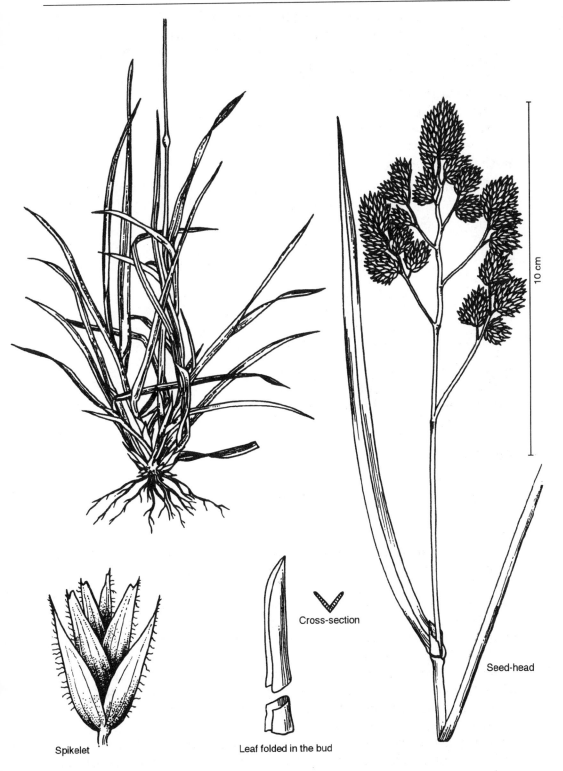

Figure 2.2 Cocksfoot (*Dactylis glomerata*)

Porto's autumn and late spring production is better than that of Currie, but it is not drought tolerant and should not be sown in areas where the rainfall is less than 500 mm (20") per annum. Other northern varieties include Cressy, Potomac and Danish.

Mediterranean types
Currie cocksfoot
Brignoles
Berber
Kasbah

These cultivars have been developed from plants collected in the Mediterranean region where the climate is similar to parts of South and Western Australia. They grow well on well drained soils in districts where the rainfall is as low as 400 mm (16") per annum, their summer dormancy giving them the ability to withstand long hot summers. These cocksfoot cultivars persist better than ryegrass on light sandy soils.

Currie cocksfoot was selected from seed collected in Algeria in 1937, and after years of testing in Western Australia by the CSIRO and the Western Australia Department of Agriculture was certified and released in 1962.

Currie makes most of its growth during autumn, winter and spring, but it also has the ability to respond to summer rains. It is an excellent grass to sow in mixed dryland pastures on well drained soils in 400 mm (16") rainfall areas.

Although best results are obtained by sowing it in either autumn or spring into a well prepared seedbed, it does possess sufficient seedling vigour to enable good results when sown with a sod seeder. It should be sown at a rate of 2–5 kg/ha.

Kikuyu
Pennisetum clandestinum

Kikuyu is a perennial grass native to central and eastern Africa. It was first introduced to Australia in about 1920, and it is now common in both tropical and temperate regions. It has also been extensively used in race courses, caravan parks, camping grounds and playing fields where it is probably the most hard-wearing of all turf species.

Kikuyu is a very vigorous summer growing perennial that is winter dormant in cooler climatic zones. It has a creeping growth habit, producing extensive rhizomes and stolons, and can be distinguished by its hairy leaves, which are folded in the bud, are without auricles, and have a ligule consisting of fine white hairs about 1 mm long.

As a pasture grass it can be used on well drained fertile soils in districts where the rainfall exceeds 600 mm (24") per annum. If grown on fertile soils and irrigated, it is capable of producing high yields of palatable forage material with a crude protein content as high as 20 per cent, in new growth.

Kikuyu pastures are common in coastal areas of eastern Australia where

Pasture grasses

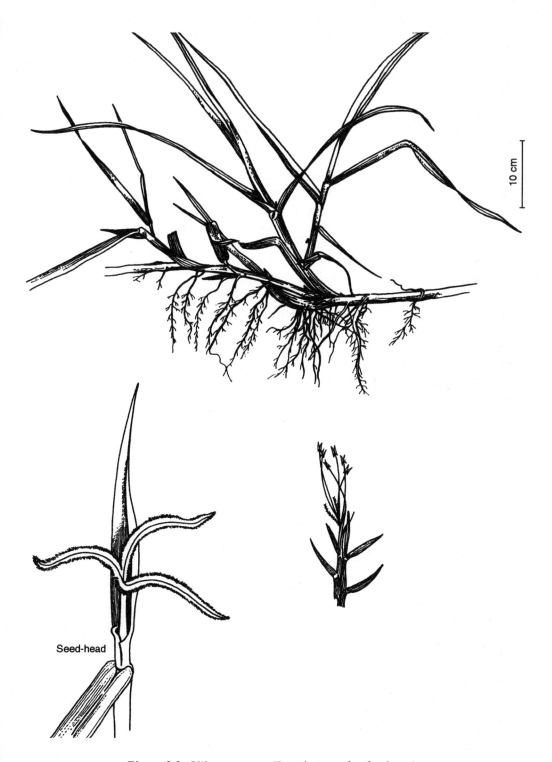

Figure 2.3 Kikuyu grass (*Pennisetum clandestinum*)

they provide the bulk of the summer feed requirements for livestock. In autumn the kikuyu pastures are sod seeded with winter growing species such as oats or ryegrass. Kikuyu must be kept short to be palatable and digestible to livestock. If nitrogen fertiliser is applied and sufficient moisture is available, kikuyu pastures can produce enormous quantities of dry matter and carry high stocking rates.

Because of its vigorous creeping growth habit, once established it is capable of withstanding very heavy grazing for long periods, a characteristic which makes it especially suitable for intensive grazing such as required by dairy farmers.

It should not be sown in areas that are likely to be used for cropping, as once established it is difficult to eradicate, although it can be controlled with the herbicide Glyphosate applied during the growing season.

Common kikuyu seldom sets seed and has to be established by planting runners in the spring.

Certified seed of the Whittet kikuyu cultivar first became available in Australia in 1972 following the introduction of the cultivar by the New South Wales Department of Agriculture from Kenya in 1960.

It is free seeding and grows slightly taller than common kikuyu.

It should be sown in spring, once ground temperatures have reached 20°C, into a fine well prepared seedbed. The seed should be drilled at 1–1.5 cm depth at 1–3 kg/ha, and then the ground should be compacted by lightly rolling for best results.

Paspalum
Paspalum dilatatum

A native of South American countries, Brazil, Agentina and Uruguay, paspalum was introduced into Australia in the 1870s by the German botanist Baron von Mueller.

It is a summer growing perennial which spreads by short rhizomes and forms dense compact clumps, often at the expense of other pasture varieties.

Paspalum can be distinguished by its broad dark green leaves that are usually puckered along the edges and often tinted purple. The young shoots are rolled in the bud and flattened and the stems are oval shaped.

It does best on fertile heavy soils and can be grown in dryland pastures in districts where the rainfall is 600 mm (24") per annum or under irrigation, but because of its lack of winter production and the fact that it tends to crowd out other pasture species, other perennial grasses are generally preferred to paspalum. Once established, like kikuyu, it will stand up to intense grazing pressure.

It should be sown in spring by sowing into a prepared seedbed or sod seeding at a rate of 1–4 kg/ha.

Once established, grazing pressure should be maintained to prevent it from flowering as the flowers are subject to infection by ergot fungi that are toxic to cattle, horses and sheep.

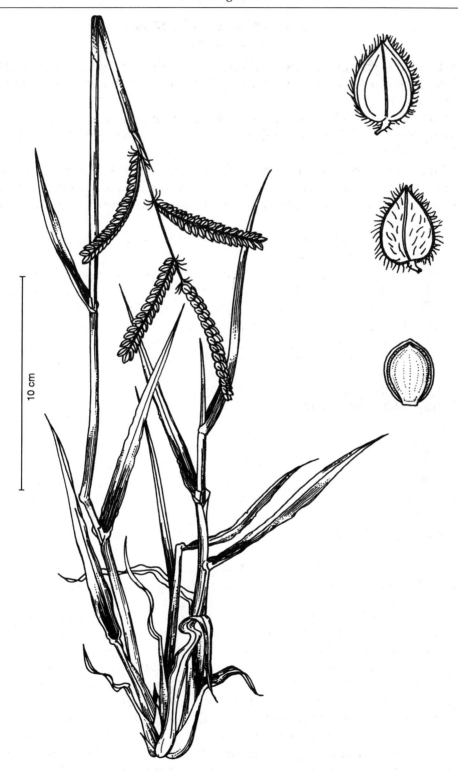

Figure 2.4 Paspalum (*Paspalum dilatatum*)

Perennial veldt grass
Ehrharta calycina

Native to South Africa, perennial veldt is thought to have been accidentally introduced to Australia earlier this century. It is now naturalised in southern Australia.

It is an extremely drought tolerant grass that will persist on poor, relatively infertile, light sandy soils in districts where the annual rainfall is as low as 350 mm (14").

It grows in erect tussocks and can be recognised by its dull green foliage. The leaves are nearly hairless, and are crinkled along one margin and rolled in the bud. They have white transparent ligules, and auricles that are often purple.

Perennial veldt produces best during autumn and spring, but it will also respond to summer rains.

It is an extremely useful grass to sow in dryland pastures on light soils in the low rainfall areas. Other species of perennial grasses are more productive and are better suited to the higher 500 mm (20") rainfall districts. It does not establish well on hard setting soils.

Veldt should be sown into a prepared seedbed at 0.5–2 kg/ha in early spring and allowed to establish well before grazing.

Grazing pressure should always be controlled to prevent this palatable grass from being eaten out.

The Mission perennial veldt cultivar was bred at the University of California Agricultural Experimental Station in 1950 from seed provided by the CSIRO, Perth, which was gathered from seed retaining plants discovered near Guildford, Western Australia. It was first certified by the Western Australia Department of Agriculture in 1968. It has the advantage over common perennial veldt of not shedding its seed, which makes reaping easier.

Besides being used for pastures, perennial veldt is also useful for stabilising and controlling light sandy soils to prevent wind erosion.

Phalaris
Phalaris aquatica, previously known as *Phalaris tuberosa*

Phalaris is native to the Mediterranean region and the cultivar Australian was selected from seed imported by the Toowoomba Botanic Gardens from the Agricultural Department of New York State, USA in 1884.

It was selected because of its vigorous growth, ability to withstand droughts and waterlogging and its ability to spread naturally, characteristics which make phalaris such a popular perennial grass today.

Phalaris is a densely tufted perennial grass with rigid stems up to one metre tall each with a knee-bend and bulbous swelling at the base. The leaves are hairless and greyish-green in colour, and rolled in the bud (a characteristic which is useful for distinguishing between phalaris and cocksfoot in the field as cocksfoot leaves are flattened in the bud). Phalaris

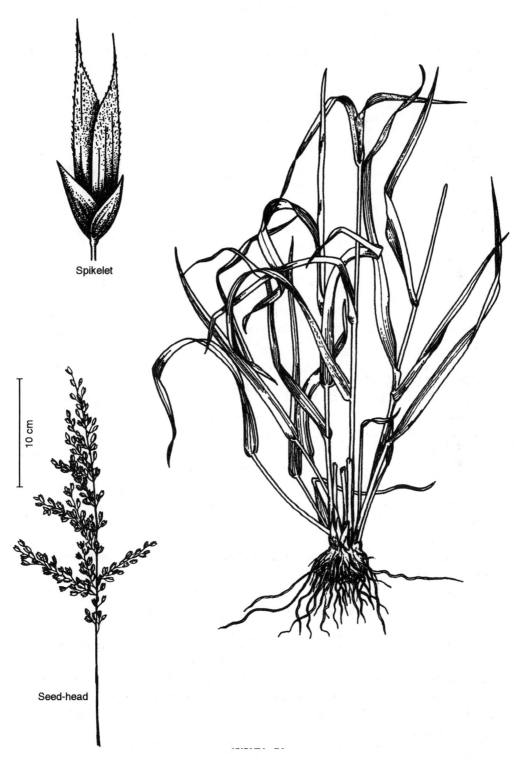

Figure 2.5 Perennial veldt grass (*Ehrharta calycina*)

leaves do not have auricles and they have prominent transparent ligules that are longer than they are wide.

Phalaris is adapted to a wide range of soil types in areas where the rainfall is 375 mm (15") per annum. Although generally slow to establish, once established, it is extremely persistent.

It is a very valuable pasture grass in dryland pastures that are too dry for perennial ryegrass to persist, and also in pastures that are irrigated only in the autumn and the spring.

Phalaris produces from autumn until late spring and persists well under heavy grazing. In fact, it has to be heavily grazed during spring to prevent it from becoming too rank and unpalatable.

All cultivars contain varying levels of toxic alkaloids, the levels of which fluctuate seasonally and under certain conditions reach levels that poison livestock. Mild cases cause phalaris staggers in sheep and severe cases cause phalaris sudden death syndrome in sheep and occasionally cattle. Phalaris staggers is a chronic condition associated with a cobalt deficiency in the phalaris diet. It can be prevented by topdressing pastures with cobalt or drenching the stock. It cannot be cured.

These alkaloids are most likely to cause problems in fresh new shoots so phalaris pastures should be grazed with care following the autumn opening rains and during the early winter, especially on phalaris dominant pastures that are recovering from a setback such as drought, fire or frost. If introducing stock into such pastures, do so when the morning dew has dried off the grass, and try and ensure that the stock are not too hungry when released into the paddock. If in doubt test the paddock with several dry sheep for a few days before turning more valuable livestock into the paddock.

Establishment of phalaris is a slow process. It does not have the seedling vigour of the ryegrass or cocksfoot cultivars, and it must be sown into a prepared seedbed without too much competition. It should be sown in autumn in drier districts or either autumn or spring in wetter areas. It should be sown 1 cm deep at the rate of 1–3 kg/ha.

Australian phalaris is recommended for 425–700 mm rainfall districts where it grows well in autumn and spring. It makes some winter growth but is summer dormant.

Siro Seedmaster phalaris. This cultivar is similar to Australian but it retains its seed better which makes harvesting easier, however it is not as persistent.

Sirocco phalaris is easier to establish than Australian phalaris and is recommended for areas where the rainfall is between 375–450 mm per annum and soils are well drained. It has a high degree of summer dormancy and produces well from autumn to spring. Summer growth following rains and cool temperatures is likely to contain dangerously high levels of toxic alkaloids.

Sirosa phalaris. This cultivar has outyielded all other perennial grasses in trials in both the south-east of South Australia and the Adelaide Hills. It

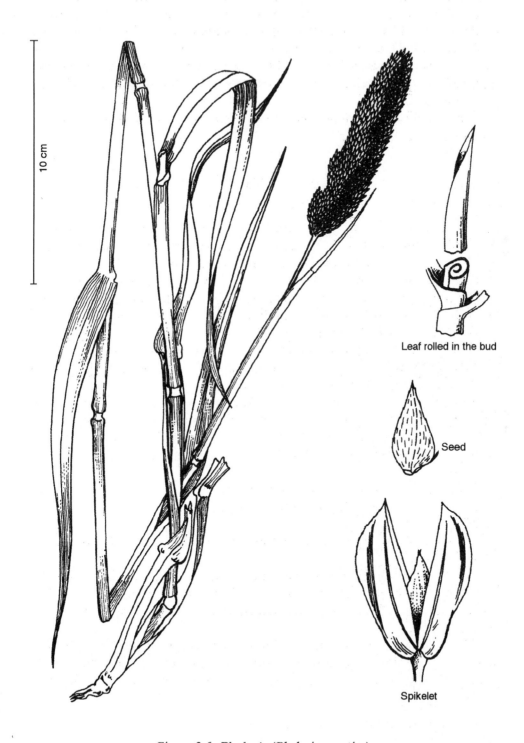

Figure 2.6 Phalaris (*Phalaris aquatica*)

has a reasonable amount of seedling vigour and is easier to establish than Australian phalaris. It is recommended for 450–700 mm rainfall districts where it produces better than Australian during winter. It also retains its seed well, and has moderately low levels of alkaloids.

Sirolan phalaris. This cultivar was developed as an alternative to Sirocco. It has lower levels of alkaloids and a lower degree of summer dormancy. It can be sown in areas where the rainfall is between 375–550 mm per annum.

Another variety is Uneta which is similar to Australian but has improved seed yield.

Prairie grass

Bromus catharticus

Prairie grass is native to Central and South America. It was introduced to Australia in the nineteenth century and has become naturalised throughout the high rainfall districts of southern and eastern Australia.

It is a tall leafy annual, biennial or short-lived perennial that grows in open clumps producing very palatable foliage. The leaves are broad, light green in colour and flattened in the bud. They can be distinguished from cocksfoot because they are slightly hairy.

Suited to irrigated pastures and dryland pastures in districts where the rainfall is 625 mm (25") per annum prairie grass thrives on well drained fertile soils, but it will not persist on poor soils.

It has the ability to grow in all seasons but produces most herbage in autumn and spring and it is particularly valuable for its winter production, especially in dairy pastures where grazing is rotated and not too heavy. Because of its palatability, it will not persist if set stocked or grazed hard for long periods.

Prairie grass can be established by either sowing into a prepared seedbed or sod seeding in autumn at rates of 5–10 kg/ha in pasture mixtures, or 20–30 kg/ha in hay crops mixtures.

The seed is susceptible to smut and should be treated with a fungicide prior to sowing.

The Grasslands Matua prairie grass cultivar was developed by the Grasslands Division of the New Zealand Department of Scientific and Industrial Research from seed provided by the CSIRO Australia. It was released in 1973.

It is a perennial that is more productive than common prairie grass, and its seed also retains its germination better. It is an excellent grass to include in dairy pastures where grazing pressure can be regulated to allow it to persist.

Figure 2.7 Prairie grass (*Bromus catharticus*)

Puccinellia
Puccinellia ciliata

Puccinellia is a densely tufted wiry perennial grass that is extremely salt tolerant. A native of the Mediterranean region, puccinellia was introduced into Australia from Turkey in 1951 by the CSIRO.

The cultivar Menemen gets its name from the Turkish town near where it was selected.

Its foliage is grey-green in colour. The inrolled to flat leaves are hairless, up to 35 cm long and taper to a point. Although it tends to grow upright in a sward, well spread plants tend to grow more prostrate.

It is essentially a pioneer plant for the reclamation of salty land. It will establish and persist on bare salt pans that flood in winter and dry out in summer, so long as subsurface moisture is available. It makes useful autumn to spring growth providing limited grazing on otherwise unproductive areas.

The seed should be sown at 1–3 kg/ha in autumn by drilling or sod seeding the seed into salt affected areas, or by broadcasting the seed mixed with fertiliser onto the areas, and then either harrowing it in or "pugging" it into the soil with livestock.

As the seed is extremely small (5 million seeds/kg) it should only be burried 0.5 cm deep.

Establishment is a slow process so the sown area should not be grazed during the first year. Once established it will withstand moderate grazing in autumn, winter and spring, but it should be allowed to set seed in the early summer if stands are sparse to enable them to thicken up from self-sown seed.

Self regenerating annual ryegrasses
Lolium rigidum

Annual ryegrass like the other ryegrass species is native to the temperate regions of Europe. It was introduced to Australia with the First Fleet, and is now widespread in all temperate regions of Australia.

Previously known as Wimmera or Early Merredin this species has bright shiny green narrow leaves and the upright stems have a reddish tinge at the base. Annual ryegrass can be distinguished from perennial ryegrass by the length of the single lower husk enclosing the spikelets on the seedhead. On annual ryegrass this lower husk is about as long as the spikelet it encloses, but on perennial ryegrass the husk is only half the length of the spikelet.

It germinates after the opening rains in autumn and produces high quality forage material in autumn and winter and especially early spring. It is normally sown in association with suitable annual medics or subterranean clovers in districts where the rainfall is between 250–650 mm (10–26") per annum on a wide range of soil types.

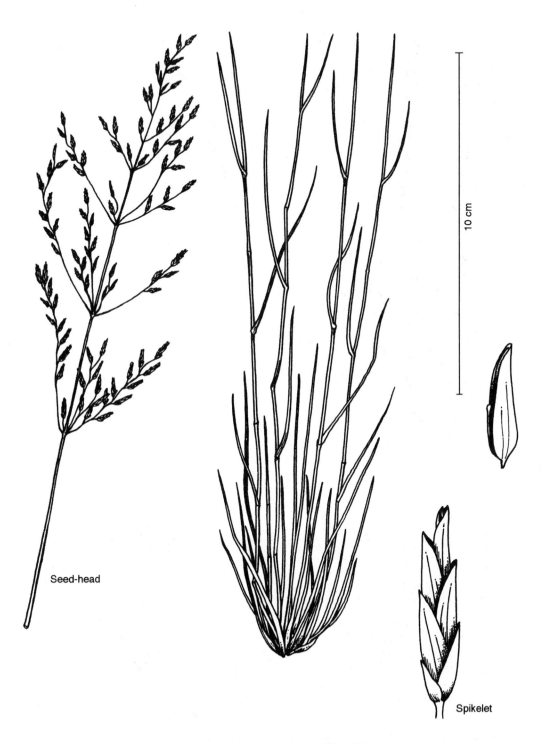

Figure 2.8 Puccinellia (*Puccinellia ciliata*)

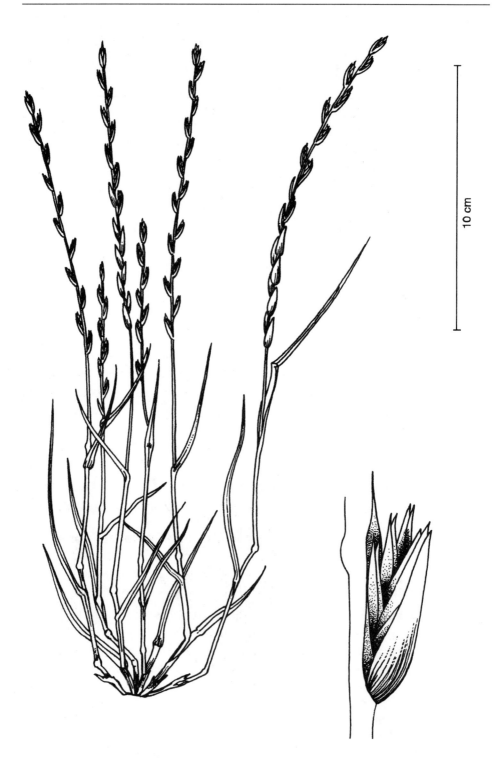

Figure 2.9 Annual ryegrass (*Lolium rigidum*)

Digestibility of annual ryegrass declines rapidly after flowering, and because of the risk of the seedheads becoming toxic, annual ryegrass pastures should either be grazed out, desiccated with a herbicide such as Paraquat or Glyphosate, cut for hay, or topped when the first heads begin to emerge. This is particularly important, because in many districts, annual ryegrass has become, or is becoming, resistant to selective grass herbicides.

Annual ryegrass is easy to establish by either sowing into a prepared seedbed or sod seeding, and, because it is free-seeding, it regenerates reliably.

Although very attractive to livestock it becomes a weed in cereal crops so management should aim to prevent it from seeding in the spring prior to the cropping year.

Guard is a new annual ryegrass cultivar that is resistant to ryegrass toxicity and susceptible to herbicides.

Short-lived Italian ryegrass
Lolium multiflorum

This annual or biennial ryegrass is native to the temperate regions of Europe and northern Asia.

It is quick to establish, extremely productive for short term pastures or hay, producing highly nutritious feed especially on fertile soils in 650 mm (26") rainfall districts, or under irrigation. Being resistant to cold conditions and frost, it produces very well during winter.

Grasslands Paroa Italian ryegrass
This cultivar was developed by the Grasslands Division of the New Zealand Department of Scientific and Industrial Research in 1934. Until 1964 it was called New Zealand Italian ryegrass, when the name was changed to Grasslands Paroa Italian ryegrass.

Its winter production is greater than that of other cultivars and should be sown at 10–20 kg/ha with red clover or shaftal clover for short term pastures for grazing or hay, or at 4–6 kg/ha in longer term pastures.

Other varieties include Midman and Concord.

Grasslands Manawa (hybrid)
This short rotation ryegrass persists for three to four years producing extremely well on fertile soils under irrigation, or in 700 mm (28") rainfall areas during autumn, winter and spring. Like Paroa it was bred by the Grasslands Division of the D.S.I.R., New Zealand. It was formerly called H1, or Short Rotation ryegrass.

Grasslands Tama (Westerwolds)
This extremely vigorous tetraploid biennial is largely replacing Italian ryegrass in some parts of Australia in short term irrigated pastures and dryland pastures in 650 mm (26") rainfall districts.

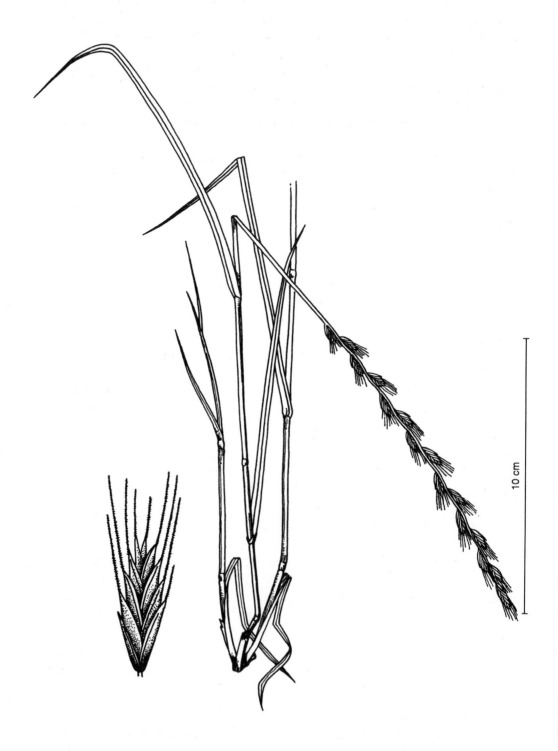

Figure 2.10 Italian ryegrass (*Lolium multiflorum*)

If sown early in the autumn in time to establish before the onset of winter it produces well in late autumn throughout winter, and spring. It is often sown with Shaftal clover (8 parts Tama:2 parts Shaftal) for winter and spring grazing and cutting for hay.

It should be sown 1 cm deep into a well prepared seedbed or sod seeded into existing pastures to boost winter production.

Grasslands Moata Tetraploid Italian ryegrass

This is another very promising short rotation ryegrass which in New Zealand has produced better than Paroa in winter and Tama in spring. It could be used as an alternative to Tama.

Other tetraploid ryegrasses include Tetila, Richmond and Barcoo.

Perennial ryegrass
Lolium perenne

Native to Europe and Asia perennial ryegrass is the most widely used species of grass sown in the high rainfall regions of temperate Australia.

Compared to other perennial species the ryegrasses are easy to establish either by sowing into a prepared seedbed or by sod seeding and they are suited to 550 mm (22") rainfall districts and a wide range of well drained soils, producing best on rich fertile loams, and persisting better on the medium to heavy rather than light soils.

Perennial ryegrass is a densely tufted grass with bright dark green leaves that are shiny underneath. The foliage produced from early autumn until spring is very palatable to livestock and has a high nutritive value.

It can be sown in autumn or spring at rates of 2–10 kg/ha depending on the mixture and rainfall. The seed should not be sown more than 1.5 cm deep. Young pastures should not be grazed hard until well established especially on light soils as stock, particularly cattle, tend to pull out young plants by their roots thinning the pasture unnecessarily. Once established though, perennial ryegrass will persist well if heavily grazed.

Perennial ryegrass pastures sometimes cause ryegrass staggers.

This condition is most likely to occur following the autumn break when there is little else for stock to eat. It is non-fatal and stock should be moved quietly to another paddock with alternative feed and water close by. They will recover within a couple of days.

Victorian Perennial ryegrass

This cultivar is the most widely sown perennial grass in South Australia. It is suited to dryland pastures where the rainfall is 550 mm (22") per annum. It establishes easily and produces best in autumn and spring.

Grasslands Ruanui

This cultivar is suited to irrigated perennial pastures and dryland pastures where the rainfall is 650 mm (26") per annum. It has a similar seasonal production pattern to Victorian, but it is not as drought tolerant.

Figure 2.11 Perennial ryegrass (*Lolium perenne*)

Grasslands Nui

Nui is more drought tolerant than Ruanui but not as tolerant as Victorian. It can be sown in place of Victorian in districts where the rainfall is 600 mm (24") per annum or under irrigation. Its autumn production is better than Victorian in dryland pastures and under irrigation its production is greater than that of Ruanui in spring, summer and autumn.

Other varieties include Kangaroo Valley, Ellett, Yatsyn and Brumby.

Tall fescue
Festuca arundinacea

Tall fescue is a deep rooted perennial grass native to the temperate regions of Europe and Asia. It was introduced to Australia in the early days of pasture development.

It grows in upright clumps and has shiny dark green foliage that is rather coarse. The leaves are stiff and erect when short, but droop as they become longer. They are thicker than the leaves of phalaris, cocksfoot and ryegrass with deep veins on the upper leaf surface and a smooth lower leaf surface.

Tall fescues are suited to irrigated pastures and dryland pastures in districts where the rainfall is 600 mm (24") per annum.

Adapted to a wide range of soil types, tall fescue will tolerate poor drainage and moderately salty conditions, but produces best on rich fertile loams.

Under irrigated conditions, tall fescues will grow all year, producing most herbage in autumn and spring, and in dryland pastures produce from autumn until late spring and respond to summer rains if they should fall.

They should be established by sowing in autumn or spring preferably into a prepared seedbed, but satisfactory results can be obtained by sod seeding. The seed should be sown 1 cm deep on heavy soils and up to 2 cm deep on light soils at 4–6 kg/ha.

Establishment of tall fescue is slow and grazing should be restricted during the first year, but once fully established it should be grazed with a high stocking rate to prevent it from becoming too clumpy, tall, rank and unpalatable. If this does happen it should be slashed and then heavily grazed with sheep.

It is not a replacement for phalaris but in high rainfall areas it will help to alleviate the problems associated with grazing phalaris dominant pastures.

Demeter fescue

This cultivar was introduced to Australia by the CSIRO from Morocco in 1931. It out-yields Australian phalaris in summer, autumn and winter, and is a very useful addition to dryland pastures. Most dairy farmers prefer the more palatable, easier to manage ryegrasses for irrigated pastures.

Epic fescue

This variety is claimed to be higher yielding than Demeter, particularly in autumn and winter.

Figure 2.12 Tall fescue (*Festuca arundinacea*)

All Triumph
A variety with improved seeding vigour and winter growth compared to Demeter.

Tall wheatgrass
Thinopyrum elongatum

Tall wheatgrass is a native of central Asia and Asia Minor. It was first introduced to Australia by the CSIRO in 1935.

It is a tall tufted perennial that is salt tolerant and adapted to marshy situations, and areas subjected to salt water flooding.

Its foliage is dark grey-green to blue-green in colour and rather coarse. It grows up to one metre tall making most growth during spring, summer and autumn.

Wheatgrass can be grown in areas where the rainfall is 450 mm (18") per annum on salty soils that are permanently wet or that dry out in summer. It should be sown into a prepared seedbed in early spring at 4–6 kg/ha for best results, and should not be grazed until well established.

Largo tall wheatgrass, an American cultivar has been most useful in pastures near the Coorong in the upper south-east of South Australia and salt affected land around Warooka on Yorke Peninsula and Edillilie on Eyre Peninsula.

It is late maturing and responds well to summer rains.

Tyrrell is another recommended variety.

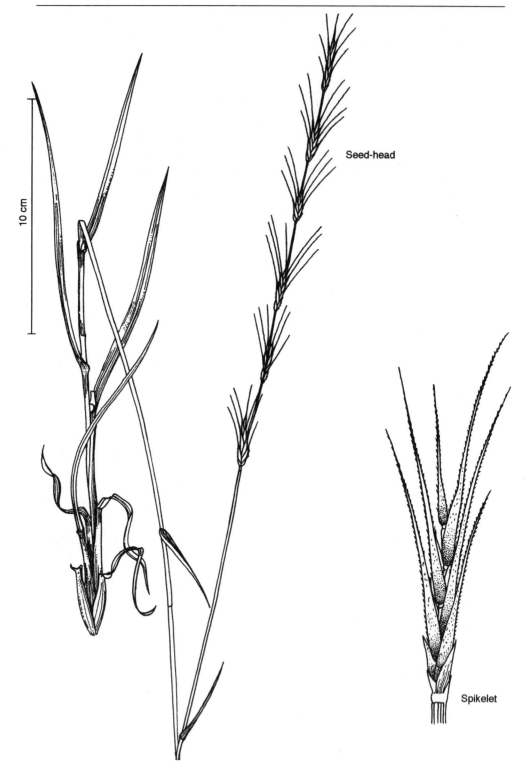

Figure 2.13 Tall wheatgrass (*Thinopyrum elongatum*)

3

Pasture legumes

Legumes are plants that belong to the family Fabaceae. The family includes pasture plants such as clovers, medics and lucerne, and grain crops such as peas, beans and lupins.

The legume content of pastures is extremely important. Besides providing a source of very nutritious and high protein forage material, legumes play the vital role of nitrogen fixation.

With the aid of rhizobium bacteria from the soil, legumes obtain nitrogen gas from the air and convert it into nitrogenous compounds. These nitrogenous compounds contribute to the fertility of the soil as the plant or parts of it die and decompose. Eventually the decomposed material will form nitrates which are the main forms of nitrogen absorbed by plants. Sustainable cropping systems rely heavily on the amount of nitrogen supplied to the soil by previous legume crops or pastures.

This process takes place when the rhizobium bacteria penetrate the roots of legumes forming small pale pink nodules on the roots. Within the nodules the bacteria is fed by the plant and at the same time excretes compounds providing the plant with nitrogenous fertiliser.

Good healthy stands of legumes are capable of fixing between 50 and 100 kg of nitrogen per hectare each year.

Other benefits to pastures and overall property management flow on from the improved nitrogen fertility levels gained from the legumes.

They include:
- Increased pasture production
- Improved nutritive value of pastures
- Higher yields of hay and silage
- Higher stock carrying capacity
- Higher yields from following crops
- Improved soil structure

❋ Reduced incidence of cereal root disease in crops following pure legume pastures
❋ Improved financial returns.

Possible disadvantages that could result from grazing some legumes, particularly if the pastures are not properly managed are:

❋ Infertility which is caused by species with a high level of the compound formononetin, which is similar to the hormone oestrogen.
❋ Increased incidence of bloat in cattle. Bloat is a build up of gas in the rumen (stomach) and if not treated promptly it can result in death.

Both these problems are most likely to occur when legumes dominate the pasture.

The benefits to be gained from legumes, however, far outweigh the disadvantages.

Amos Howard first discovered one of our most important annual legumes, subterranean clover, near Mount Barker in the Adelaide Hills in 1889. Howard recognised the benefits of this plant and ever since then the use of legumes has been promoted in pastures.

The nitrogen fixing ability of legumes is of inestimable value, especially since other forms of nitrogenous fertilisers are often uneconomical to use on pastures. This fact is aggravated by the unstable nature of the nitrogenous fertilisers which often results in losses of nitrogen due to evaporation and leaching before it can be used by plants.

There is a wide range of legumes sown in pastures in Australia. They include perennial species such as lucerne, white clovers and strawberry clover, and a range of annual species, the most important of which are the subterranean clovers and medics. In the tropical and subtropical zones leguminous species such as sirato, desmodium, lablab, stylo and centro are becoming an important component of pastures. On a broader basis small leguminous trees and shrubs are also being used for grazing. These include plants such as tagasaste or tree lucerne in temperate areas and leucaena in tropical areas.

All these different species and their cultivars have different growing requirements and different growth characteristics. Growers should make sure that they do their homework before buying seed to ensure that the species and cultivars that they sow are the most suitable for the area in which they are to be sown and for the purpose for which they are intended.

With all legumes it is necessary to consider soils and climatic conditions before sowing, but it is particularly important when considering the two main groups of annual pasture legumes, subterranean clovers and medics

The medics, species of the genus *Medicago*, usually prefer neutral to alkaline soils.

Clovers, species of the genus *Trifolium*, do best on slightly acid soils.

Apart from preferences for soil alkalinity or acidity most legumes prefer well drained soils, however, some including two subterranean clover

Pasture legumes

Figure 3.1 Medic

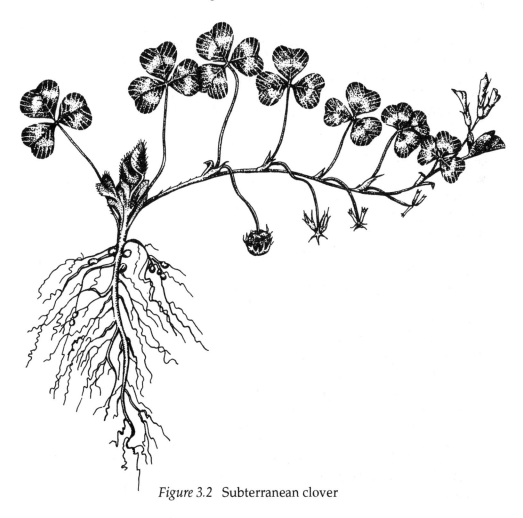

Figure 3.2 Subterranean clover

cultivars Trikkala and Yarloop, and strawberry clover do well on soils that become waterlogged in winter.

Soil fertility is extremely important. No legume will produce worthwhile yields of herbage or nitrogen on soils that lack phosphorus.

Annual legumes also require a growing season that is long enough for them to complete their life cycle: to germinate, grow, flower and mature seed. They vary greatly in the time required to do this, from as little as 80 days for some medics, to as long as 150 days for some clovers.

In late districts, late maturing annuals should be sown to take advantage of the longer growing season to produce more. In early districts the same cultivars would fail because they would not have sufficient time to set seed so earlier maturing cultivars, although less productive, must be sown.

Annual legumes are mainly sown in either short term pastures in rotation with cereal crops in the warm temperate zone or in association with perennials in non-irrigated pastures in the high rainfall areas.

Perennial legumes, on the other hand, are sown with perennial grasses in non-irrigated pastures in high rainfall districts and in irrigated pastures.

Lucerne, a perennial, is a notable exception. Lucerne is mainly sown in pure stands in both non-irrigated and irrigated situations and can be grown in districts where the rainfall is as low as 350 mm (14") per annum as well as in the high rainfall zone.

Inoculation

Rhizobium bacteria often occur naturally in the soil, but appropriate strains for particular species are not always present. It is therefore a wise precaution when sowing legumes to inoculate the seed, particularly when sowing new land or introducing a new species of legume to old country, for without the bacteria the legumes are unable to perform their nitrogen fixation role.

The appropriate strains of bacteria are shown in Table 3.1 at the top of the following page and are available from leading seed merchants.

To inoculate legume seed the following items are needed:

- ❈ Seed of a known species
- ❈ Appropriate bacterial culture
- ❈ Adhesive
- ❈ Finely ground agricultural lime
- ❈ Mixing equipment such as a concrete mixer, bucket, or even just a clear space on a clean concrete floor and a shovel
- ❈ Fine sieve.

The steps to follow to inoculate seed are:

- ❈ Weigh the seed and place a known manageable quantity into the mixer.
- ❈ Add sufficient glue to dampen the seed. This must be done while mixing to ensure that the seed does not become too wet.

Pasture legumes

Table 3.1 Bacterial culture groups pasture legumes

Legumes inoculated	Rhizobium group	Size* kg
Lucerne and Medics	Group A	50
Red, Strawberry, Shaftal, White Clover	Group B	25
Subterranean Clover, Crimson Clover	Group C	50
Lotus pedunculatus (L. major)	Group D	25
Pea, Vetch, Tares, Faba Bean, Lentil	Group E	100
French, Climbing, Navy Beans	Group F	100
Lupin	Group G	100
Seredella	Group G	50
Soybean	Group H	100
Cowpea, Peanut, Velvet Bean	Group I	100
Siratro, Phasey Bean, Puero, Calopo	Group I	50
Glycine, Stylo (except Verano & Oxley)	Group I	25
Lablab	Group J	100
Axillaris	Group J	25
Kenya White Clover	Group K	25
Lotononis	Group L	10
Mung Bean	Group M	100
Centro	Centro	100
Chickpea	Chickpea	100
Desmodium	Desmodium	25
Leucaena	Leucaena	100
Oxley Stylo	Oxley	25
Verano Stylo	Verano	25

*Each standard pack inoculates the weight of seed shown below.

- Add sufficient bacterial culture to treat the seed and mix thoroughly. The amount of bacterial culture required can be calculated from the inoculation chart recommendations.
- Once the seed has been thoroughly covered with the culture, microfine lime should be added gradually while still mixing, until the seed is covered. Generally approximately 10–15 kg of lime are needed to coat 50 kg of clover, lucerne or medic seed.
- Sieve the seed which has been inoculated and lime coated to remove any lumps of seeds that have stuck together. It may be necessary to add more lime and remix if too many lumps are present.
- Sow the seed as soon as possible. Remember the bacterial culture is susceptible to heat and will be killed if the treated seed is left standing about for too long before sowing.
- Unused bacterial culture must be stored in a cool place, preferably in a refrigerator.

It is good policy to only treat enough seed for a day's work before starting seeding each day. If purchased seed is already inoculated, it must be used promptly.

Once the legumes have established in order to determine if the bacteria are working and fixing nitrogen a few plants should be dug up and the soil carefully washed from their roots. If the nodules on the roots are plentiful, soft and pale pink in colour then the bacteria are healthy and doing their job. If, however, there are not many nodules present and those that have formed are hard and either white or brown in colour then something is wrong and nitrogen fixation will be poor.

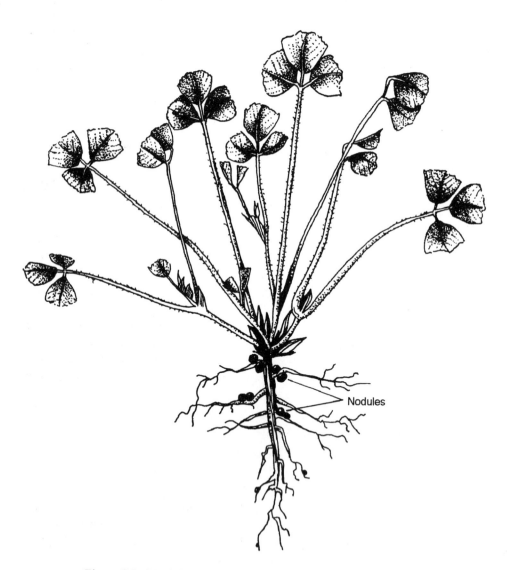

Figure 3.3 Nodules containing nitrogen fixing bacteria on the roots of a legume plant

Subterranean clovers
Trifolium species

Subterranean clover (commonly called sub clover) is the most important annual legume sown in southern Australia. It is a free seeding, self regenerating plant with a prostrate growth habit which buries its seed, a characteristic that is responsible for the name subterranean (see Figure 3.2).

The species originated in the Mediterranean region, parts of western Europe, southern England and Ireland. Probably it was accidentally introduced to Australia in hay as early as the 1830s but its potential as a pasture plant was not recognised until it was discovered by Amos Howard growing on his property near Mount Barker in 1899.

Following Howard's discovery the first commercial sales of subterranean clover seed were made in 1906 and since then, and especially since the development and accepted use of phosphatic fertilisers in the 1920s, its use has continued to grow.

Sub clovers should be sown to make an evenly balanced pasture of grasses and legumes. This means the mixture should consist of approximately two-thirds grass to one-third legume seed, by weight.

They prefer soils that are neutral to slightly acid and an annual rainfall in excess of 400 mm (16").

Sub clovers establish well if they are either sod seeded or sown into a prepared seed bed and can also be established by broadcasting.

Grazing during the first year after sowing should be light and regulated to enable flowering and a good seed set. First year pastures should not be cut for hay. Following the establishment year, sub clovers will withstand heavy grazing and their prostrate growth habit and seed burying ability, as well as their ability to produce hard seed, will enable them to persist.

Pests such as red-legged earthmite and lucerne flea can cause extensive damage, especially to seedling plants in the autumn, and care should be taken to avoid this, spraying if necessary.

The commercial sub clovers belong to one of three closely related species, *Trifolium subterraneum*, which contains the most cultivars, *T. yanninicum* or *T. brachycalycinum*.

T. yanninicum cultivars include Trikkala and Yarloop that are white seeded and well suited to waterlogged soils.

T. brachycalycinum includes the cultivar Clare that is able to grow in soils that are slightly alkaline.

Cultivar identification

Because sub clover is normally self fertilised in the field there is a marked uniformity of characters within each cultivar and wide variations between cultivars. They can be identified by differences in appearance as shown in Figure 3.4.

Remember it is important to identify the cultivars that are doing well in particular areas so that you can use them for seeding in similar situations

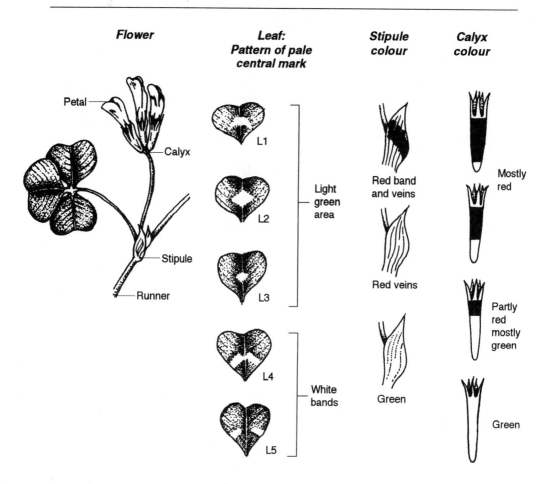

Figure 3.4 Parts of a sub clover plant used for identification

and also make enquiries with the Department of Agriculture and seed merchants to find out if there are any better cultivars to use in the areas.

Unless they are being sown for pure seed production pastures should be sown to mixtures of at least two, preferably three different cultivars with differing maturities. This will allow for varying seasonal conditions and variations in soil type and possibly drainage to ensure the best production.

The most important characteristics to consider when making your choice are:

- Maturity — the length of time that each cultivar takes to reach maturity and set seed. You must choose ones that will set seed reliably in your district. This depends mainly on the annual rainfall pattern.
- Soil type — well drained neutral to slightly acid soils are preferred. Few clovers will withstand waterlogging. Trikkala is the cultivar to use in poorly drained areas, and only Clare will tolerate and produce well on slightly alkaline soils.

Table 3.2 Varietal characteristics of subterranean clover cultivars

Variety	Flowering starts	Seeds formed	Leaf pattern of pale central mark	Stipule colour	Runner hairiness	Petal colour	Calyx colour	Seed colour	Other distinctive features
Nungarin	early August	late September	L1	Light red veins	Hairy	White-pink	Mostly red	Black	Leaflets distinctly triangular
Geraldton	mid-late August	early October	L1	Red veins	Very hairy	White-pink	Mostly red	Black	Corolla sometimes distinctly pinkish
Dwalganup	mid-late August	mid-late October	L3 L5	Red veins	Very hairy	White-pink	Mostly green	Black	No flecking on leaves
Dalkeith	late August	mid-late October	L3 L5	Red veins	Very hairy	White-pink	Mostly green	Black	
Daliak	early September	mid-late October	L3	Red band and veins	Hairy	White-pink	Purple-red	Black	Calyx distinct purple-red
Yarloop	early September	late October	L5	Red band and veins	Hairless	White	Green	White	Midrib often brownish colour
Trikkala	mid September	late October	L2 L4	Pink veins	Hairless	White	Green	White	
Seaton Park	mid September	late October	L2 L4	Green	Hairy	White-pink	Green	Black	
Esperance	mid September	late October	L3		Hairy	Mostly red	Black		
Woogenellup	mid-late September	early-mid November	L1	Red veins red band	Hairless	White	Green	Black	Stipules larger than normal
Clare	mid-late September	mid November	L2 L4	Red band and veins	Hairless	White	Green	Black	Distinct purple colour below pale mark
Mt Barker	late September	late November	L2	Red band and veins	Hairy	White-pink	Mostly red	Black	Distinct red calyx band
Meteora	late September	late November							
Larisa	late September	late November	L2 L4	Pink veins	Hairless	White	Green	White	

Table 3.3 Characteristics of subterranean clover species

Cultivar	Species	Maturity	Height of pasture	Minimum rainfall	Formonon-etin levels	Clover scorch (Kabatiella)	Soil types	Farming system
Nungarin	sub	early August	low	350 mm	low	1	sands to loams	Used in crop rotations in cereal systems, and in permanent pastures
Geraldton	sub	late August	low	400 mm	moderate to high	1	sands to loams	Shows good hard seededness and burr burial, but can cause infertility in ewes.
Dwalganup	sub	mid-August	low	400 mm	high	1	sands to loams	Not recommended because it causes ewe infertility. Replace with Daliak or Dalkeith in wetter areas, and Nungarin in drier.
Dalkeith	sub	late August	low	400 mm	low	4	sands to loams	Promising new cultivar, use as a replacement for Dwalganup.
Daliak	sub	late August	low	425 mm	low	9	sands to loams	Could be used in a mixture with Nungarin for heavier soils.
Yarloop	yann	late August	tall	450 mm	high	1	low-lying soils subject to waterlogging	Tolerates waterlogging for long periods but not recommended because it causes ewe infertility. White-seeded with good winter growth. Replace with Trikkala.
Seaton Park	sub	early September	medium	450 mm	low	1	sands to loams	A reasonably vigorous strain that persists in areas of moderate rainfall. Can be used as a replacement for Dwalganup. Trikkala preferred.
Trikkala	yann	early September	medium	450 mm	low	4	loams to lowlying soils subject to waterlogging	A white-seeded type with good winter growth. Use as a replacement for Yarloop and Seaton Park.
Esperance	sub	mid-September	low	450 mm	low to moderate	9	sands to loams	Bred for resistance to clover scorch.
Woogenellup	sub	mid-September	very tall	450 mm	low	1	sands to loams	A vigorous and very productive strain used in grazing systems. Susceptible to clover scorch. Trikkala preferred.
Clare	brachy	mid-September	very tall	450 mm	low	2	acid to slightly alkaline loams	Useful in the cereal zone where soils are slightly alkaline. Fairly soft-seeded and does not persist well under grazing.
Mt. Barker	sub	late September	medium	500 mm	low	6	sands to loams	Suited to a wide range of soils and conditions where rainfall exceeds 500 mm.
Meteora	sub	late September	medium	550 mm	low	4	sands to loams	Promising in late districts.
Larisa	yann	early October	low	600 mm	low	5	low-lying soils	Used in areas of poorly drained soils and high rainfall. Requires a very long season.

Species: sub — *T. subterraneum*; yann — *T. yanninicum*; brachy — *T. brachycalycinum*.
Clover scorch resistance rating; 1 — little or no resistance to clover scorch; 10 — very high resistance to clover scorch.

- ❊ Clover disease — select cultivars that contain low levels of the hormone, oestrogen formononetin, that causes infertility in ewes.
- ❊ Clover scorch or Kabatiella — choose cultivars that are resistant to this fungal disease to prevent severe losses in spring.

Other characteristics which may also have to be investigated are the degree of seed hardiness, burr burying ability, preferences for different soil types and of course seasonal growth and productivity.

Annual medics
Medicago species

Annual medics are self regenerating legumes suited to the alkaline warm temperate soils in the 250–500 mm (10–20") rainfall districts of the warm temperate zone.

The introduction of these legume pastures into crop rotations started during the late 1930s and early 1940s and has resulted in higher and more regular returns from both cereal crops and livestock.

Figure 3.5 Typical medic features

The medics that were first used in these pastures were either introduced from countries around the Mediterranean Sea or selections from those cultivars. Burr medic and Hannaford barrel medic were amongst the first medics to be used and the range grew over the years to include Jemalong, Cyprus, Harbinger, Paragosa, and Tornafield.

Unfortunately few of these cultivars are tolerant or resistant to the aphid pests (spotted alfalfa aphid, blue-green aphid and pea aphid) that were introduced to Australia in the mid 1970s. Of the medics in use at that time only Snail, Jemalong, Paragosa and Cyprus showed any tolerance.

Since then a number of new resistant cultivars have been introduced or developed and like the older ones they all have their own characteristics and grow best under certain conditions. Resistant varieties are now used wherever possible.

Medic pastures can be sown by one of three different methods:
- Drilling the seed into cereal stubbles before the opening rains
- Undersowing with the cereal crop
- Sowing into a prepared seedbed.

The first method of dry sowing into stubbles is cheap and effective, seed being scratched into the soil at the rate of 5–10 kg/ha.

The second method is sometimes convenient but risky. Besides the medics competing with the crop they may have to survive herbicide applications and uncontrolled insect attacks. Sowing rates using this method are usually light and between 2–4 kg/ha.

Sowing into a prepared weed free seed bed is by far the most effective way of establishing pure medic and grass free pastures. Seeding rates using this method should be at least 8–12 kg/ha and the seed should be sown between 1.5 cm and 3 cm deep depending on soil type.

Following establishment medic pastures should be grazed to keep them relatively short until flowering starts. Stock should then be removed to enable flowering and podding (seed set).

When the pasture has reached maturity and dried off, the dry residue, including the pods, is a valuable source of high protein feed but care should be taken not to overgraze during the summer so that good seed reserves are left to ensure regeneration. The pods from burr medic are also a source of contamination in wool from sheep grazing in it.

Even following cropping and false breaks to the season, in districts with very low rainfall most medics will regenerate reliably because they set a high proportion of hard seed.

Cultivars vary greatly and it is important to be aware of growing requirements and growth characteristics before sowing.

Like sub clovers, medics will perform best when managed well. Besides controlling grazing, growers must apply regular applications of superphosphate. At least 100 kg/ha is required annually.

Insect pests such as red-legged earthmite, lucerne flea and sitona weevil as well as the aphids must be controlled.

Weeds too, must be controlled to minimise competition and yields.

Six different species of the genus *Medicago* are currently available commercially in Australia. They are as follows:

M. littoralis (strand medic)
M. polymorpha (burr medic)
M. rugosa (gama medic)
M. scutellata (snail medic)
M. tornata (disc medic)
M. truncatulata (barrel medic)

The different species can be identified by various characteristics but the seed pods provide the easiest means of identifying the species in the field, as the shown in Table 3.4 on the next page.

Sundry annual legumes

There are a number of sundry annual legumes that can be sown in pastures in particular situations for particular purposes in southern parts of Australia.

They include balansa clover, rose clover, serradella, shaftal clover and vetch.

A number of other annual legumes such as cluster clover, and melelotus are occasionally used but their use is very limited, melelotus being restricted to salt affected areas, where it has been used as a pioneer.

Balansa clover
Trifolium balansae

Balansa clover is a recent addition to the list of pasture legumes, 1985 being the first year that a good supply of seed was available.

Introduced from Turkey, balansa clover is adapted to most soils that suit sub clover, in districts where the annual rainfall is 450 mm (18") or more. It also has the ability to thrive on soils that become waterlogged during winter and, although it does not produce as much winter growth as Trikkala sub clover, it outyields Trikkala in spring. It is not recommended for deep acid sands or alkaline soils.

One major attribute of balansa clover is its high tolerance to clover scorch (kabatiella), the fungal disease that causes severe losses in some legume pastures, particularly in spring.

Balansa clover establishes best if sown no more than 1 cm deep into a well prepared, fine, weed free seed bed at 2–3 kg/ha.

Once established balansa will stand set stocking and can also be cut for hay. It is a prolific seeder and regenerates reliability.

Paradana is the only cultivar currently available.

(See page 60, Figure 3.6, for for an illustration of this plant.)

Table 3.4 Annual medic identification table

Name	Cultivars	Stem	Leaflets
Barrel medic *M. truncatulata*	Borung, Cyprus Hannaford, Jemalong, Paraggio, Sephi	prostrate to upright; hairy	hairy, marking vary from none to brown flecks to a brown wedge (cv. Jemalong)
Burr medic *M. polymorpha*	Serena	mainly prostrate; very hairy	hairy, margin toothed
Disk medic *M. tornata*	Tornafield	prostrate; slightly hairy	unmarked; slightly hairy; toothed margin
Gama medic *M. rugosa*	Paragosa, Paraponto, Sapo	prostrate or ascending; hairy	unmarked; hairy only on lower surface; rounded at ends, margins toothed
Snail medic *M. scutellata*	Sava	prostrate to ascending; very hairy	unmarked, ovate or wedge shaped; margin toothed; hairy
Strand medic *M. littoralis*	Harbinger	mainly prostrate; hairy	unmarked; hairy on both surfaces, margin toothed

Please note that all medic cultivars can be easily distinguished from clover cultivars in the field simply by comparing the leaves. The petiolules joining the leaflets to the petiole on clover cultivars are of equal

Rose clover
Trifolium hirtum

Rose clover is a native of southern Europe and the Mediterranean region. Adapted to soils that are low in fertility where the annual rainfall is 350 mm (14") or more rose clovers can be used on soils that are too acid to grow medics and where sub clovers such as Nungarin, Geraldton and Daliak fail to persist. They have also been successfully used as pioneer legumes on deep acid sands in the higher rainfall districts where the fertility level has been too low for sub clovers (Figure 3.7 on page 61).

Rose clover should be established by sowing at rates of 2–4 kg/ha.

It grows to a height of 40–50 cm but does not bury its seed or even set it close to the ground like sub clovers and medics, so it should not be grazed once flowering starts until after the seed has matured. Its ability to set a high percentage of hard seed ensures regeneration if grazing is controlled in this way.

The cultivar Kondinin rose clover is adapted to 350–450 mm (14–18") rainfall districts. Its yields are similar to those of Nungarin sub clover and Cyprus barrel medic.

Pasture legumes

Table 3.4 (continued)

Stipule	Flowers	Seed Pods
toothed	yellow; usually 2–4 on short stalks in axils of leaves	barrel shaped burr with spines but no hooks; 4–5 close coils; Cyprus and Hannaford coil clockwise, other anticlockwise
broad, may have few small teeth near base	yellow; 1–3 on awned stalk no longer than leaf stalk	burr has straight spine; 2–5 coils
toothed	yellow; 4–15 on stalks longer than leaf stalks and growing from leaf axils	pod is flat round disc shape; spineless with 2–5 coils
toothed	yellow; 2–5 on short stalks in axils of leaves	pod is a round flat spiral disc with conspicuous radiating veins; spineless
tapering; toothed	yellow; 1–3 on awned stalk shorter than the leaf stalk	large snail-shaped spineless pos with 5–8 coils
toothed	yellow; usually 3–4 but up to 6 on short stalks in axils of leaves	similar to but smaller than burr of barrel medics with short spines and only 3–4 close coils

length but on all medic cultivars the terminal petiolule joining the centre leaflet to the petiole is longer than the two lateral petiolules joining the other two leaflets to the petiole.

Hybon rose clover is earlier maturing than Kondinin and it regenerates more reliably in drier districts. Hybon can be sown in areas where the rainfall is as low as 325 mm (13") per annum.

Serradella
Ornithopus compressus

Serradella is a native of southern France, Portugal and Spain. It is an annual herb with a spreading growth habit growing up to 40 cm high. It was introduced during the 1950s.

Although not widely used this deep rooted plant has the ability to grow on relatively infertile acid sandy soils where the annual rainfall is 500 mm (20"). It has proved to be a useful pioneer legume in these areas.

Serradella sets an extremely high percentage of hard seed which softens under prolonged exposure to alternating high and low temperatures in the field or under controlled conditions. It is best to sow seed that has been heat treated to ensure a good initial germination and establishment.

Seed should be sown at 4–8 kg/ha no more than 2 cm deep. Once established it will stand light grazing.

Figure 3.6 Balansa clover (*Trifolium balansae*)

Figure 3.7 Rose clover (*Trifolium hirtum*)

Pitman and Uniserra are the two available cultivars. They are very similar but Uniserra matures 1–2 weeks earlier than Pitman and is therefore likely to regenerate more reliably in drier areas where the annual rainfall is as low as 350 mm (14").

Other varieties include Avila which is late maturing and tolerates soils with high levels of exchangeable aluminium. Elgara, Madeira and Paros are earlier maturing and only moderately tolerant of acid soil conditions.

Shaftal clover
Trifolium resupinatum

Often referred to as Persian clover, after the country of origin, shaftal clover came to Australia in the 1960s. Its potential as a pasture plant was recognised and it is now extensively used in short term winter pastures.

Adapted to a wide range of well drained fertile soils shaftal makes exceptional winter and spring growth.

Shaftal should be sown into a fine, well prepared, weed-free seed bed early enough in the late summer-early autumn period to allow it to establish well before the onset of cold winter weather. If it is not sown early much of the benefit from its winter production will be lost.

Because it has very small seed shaftal should not be sown more than 1 cm deep and because of the shallow early seeding it is usually necessary to irrigate to ensure early establishment.

In short term winter pastures shaftal should be sown at 4–6 kg/ha when it is the only clover used or at 2–3 kg/ha if other clovers such as Paradana are also sown. The balance of such mixtures should be made up of vigorous ryegrasses such as Tama or Moata.

Pasture management

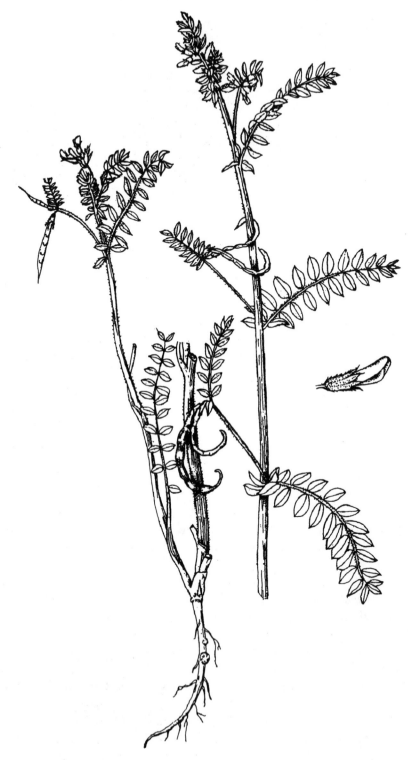

Figure 3.8 Serradella (*Ornithopus compressus*)

Figure 3.9 Shaftal clover (*Trifolium resupinatum*)

Once established shaftal will produce well until late spring. It will withstand heavy grazing and makes excellent hay. Shaftal does not seem to cause bloat to the same extent as most other clovers.

Maral and Kyambro are the only cultivars available.

Vetch
Vicia species

Vetch is another pasture introduction from southern Europe. It is a pea-like plant with a climbing growth habit adapted to a wide range of different soils including relatively infertile acid soils in 375 mm (15") rainfall districts.

Vetch is mainly used for improving soil fertility or sowing in mixtures with oats and ryegrass to cut for hay, although it can withstand grazing and regenerate well.

Vetch has relatively large seed (the size of a small pea) and establishes well even when sod seeded or broadcast onto rough seedbeds at rates of between 10–40 kg/ha depending on the mixture.

It seems to be more palatable to sheep than cattle. It should be lightly grazed throughout the winter and in spring it can either be heavily grazed or shut up for hay production.

Woolly pod vetch — *Vicia dasycarpa*

Namoi is the commonly used cultivar of this species. It has the ability to set seed under very adverse conditions, and produces well on a wide range of soils either acid or alkaline.

Purple vetch — *Vicia berghalensis*

Popany purple vetch can be used in similar situations to Namoi woolly pod vetch in districts where the annual rainfall is at least 500 mm (20"). It makes excellent hay, but is not as palatable to livestock as Namoi when it is green.

Lucerne
Medicago sativa

Lucerne is the most widely sown perennial legume in southern Australia.

Lucerne is long living plant with a deep tap root system and an upright growth habit making it suitable for growing in either dryland or irrigated pastures for both grazing and cutting for hay.

In dryland situations although production may vary greatly with seasonal conditions the plant population of lucerne stands tend to remain relatively stable unless affected by severe overgrazing, waterlogging, or attack from disease organisms or insect pests such as aphids.

Other advantages to be gained by growing lucerne are:
- ❀ Rapid early growth
- ❀ It can remain green at the end of the season and respond to summer rains which are of little or no use to other annual pasture species

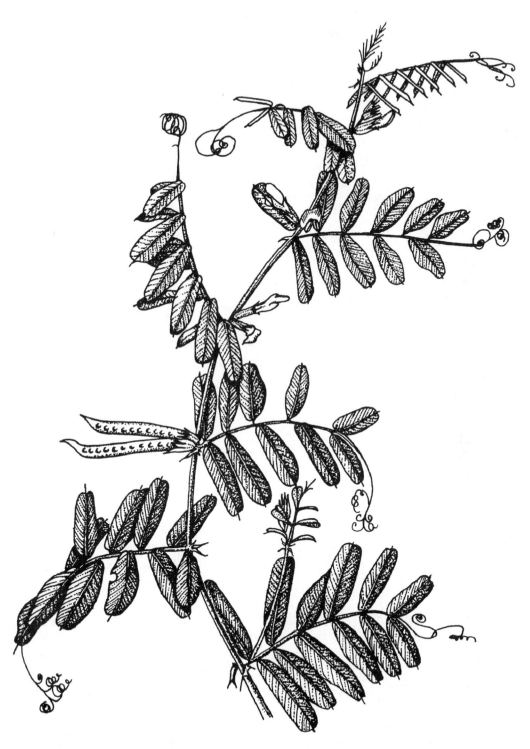

Figure 3.10 Vetch (*Vicia* spp)

Pasture management

Figure 3.11 Lucerne (*Medicago sativa*)

❁ Lucerne is extremely drought hardy and can withstand dry spells in years of irregular rainfall.

Lucerne grows well on a wide range of well drained soil types particularly deep, fertile, neutral to alkaline soils.

It will persist in districts where the annual rainfall is as low as 300 mm (12"), but grows particularly well under irrigation. It can withstand moderate levels of salinity.

The management of lucerne differs from that of other pastures, both perennial and annual. Lucerne is slow to establish and is best sown in pure stands or with a cover crop into a prepared, weed-free seedbed. First year stands need particularly careful management to allow the plants every opportunity to develop a good root system before the onset of summer. They cannot do this if overgrazed. Where the rainfall is 375 mm (15") or less, grazing during the first year is seldom advisable.

Unlike the other pastures lucerne pastures must be rotationally grazed. The plants survive best when the grazing period is short and the rest period is long. This method avoids grazing animals damaging the crowns and young shoots and maximises production.

When cut for hay lucerne should be mown at the early flowering stage. Under good conditions 6–8 cuts can be obtained annually.

Well-managed lucerne stands should persist for five years or even longer, providing good fertiliser, pest control and weed control practices are maintained with strict grazing or mowing practices.

There is a wide range of different cultivars available. All have differing growth characteristics, but since it is almost impossible to identify the different cultivars by examining their botanical features, it is essential that intending growers purchase certified seed.

The cultivars vary greatly in seasonal growth from those that are winter active, to others that are winter dormant. Remember that winter growth can only be effectively utilised by grazing.

Apart from seasonal growth other major characteristics to consider when deciding which varieties to sow are resistance to diseases such as bacterial wilt, fusarium wilt, anthracrose (colletotrichum crown and root rot), and phytophthera root rot, and most importantly aphids.

Three aphid pests attack lucerne; the spotted alfalfa aphid, the blue-green aphid and the pea aphid. Resistance to all three is important, especially the spotted alfalfa aphid which causes damage mainly in autumn and spring.

These characteristics are all summarised in Table 3.6.

Growers should consult experienced local producers and their regional office of the Department of Agriculture if in doubt as to which cultivars to sow, and for purposes other than seed production sow at rates ranging from 25 kg/ha under irrigated conditions down to 2–4 kg/ha under dryland conditions where the rainfall is as low as 300 mm (12") per annum.

New varieties of lucerne are constantly being released onto the market. Each year recommended varieties will change and it important for producers to keep up to date with information. The varieties listed in Table 3.6 may not be current and are not exhaustive.

Sundry perennial legumes

Three species of perennial legumes other than lucerne are commonly grown in southern Australia. They are in order of importance white clover, strawberry clover and red clover. All are used in irrigated pastures and white clover and strawberry clover in non-irrigated pastures in the high rainfall areas, where the rainfall is at least 600 mm (24") annually.

White clover
Trifolium repens

The different cultivars used in southern Australia are New Zealand, Haifa and Tarma. White clovers spread by rooted runners producing high quality stock feed mainly during the spring, summer and autumn.

Table 3.6 Pest resistance and growth features of lucerne cultivars

Variety	Major Factors for Selection				Other Factors		
	Late autumn winter/ growth	Spotted alfalfa aphid	Blue green aphid	Phytophthora root rot	Colletotrichum crown rot	Stem nematode	Bacterial wilt
Recommended varieties							
Winter-dormant							
Pioneer Brand 545	2	HR	S	R	LR	LR	R
Semi-dormant							
Nova	5	HR	S	MR	S	LR	R
Pioneer Brand 581	5	R	S	MR	LR	MR	R
Pioneer Brand L52 (P)	5	HR	LR	R*	LR	R*	HR*
Validor#	5	MR	S	MR	S	S	R
WL Southern Special	6	R	LR	MR	MR	LR	R
Winter-active							
Aurora	6	HR	MR	R	MR	R	-
Hunterfield	6	HR	LR	S	S	S	S
Trifecta	7	MR	LR	MR	R	LR	-
WL516	7*	HR	LR	R	MR	MR*	MR*
Highly winter-active							
Pioneer Brand 577	9	HR	S	LR	LR	S	S
Maxidor II	9	HR	LR	MR	LR	MR	MR
Sequel	9	R	LR	MR	R	S	S
WL605	9	R	LR	R	S	MR*	S*
Pioneer Brand 5929	9	R	LR	R	S	LR	-
CUF 101**	9	HR	LR	MR	S	S	S
Recommended but outclassed (An outclassed variety is one which has been superseded in most situations by a recommended variety/varieties according to trial results where available, or by broadacre observations of overall performance.)							
Siriver	9	HR	MR	S	S	S	S
Merit to be established							
Quadrella	7*	MR*	LR*	LR*	R*	LR*	-
Other varieties—not recommended							
Cimarron	4/5	LR	S	MR	MR	-	R
Hunter River	5	S	S	S	S	S	S
Falkiner	5	LR	S	MR	S	R	R

P Provisional recommendation. Made after one year but less than three years of trials or satisfactory broadacre experience.
† Varieties are listed in order or increasing late autumn winter growth from 2 (very slow growth) through 6 (moderate growth) to 9 (very active growth). Dormancy groups ;should not be considered absolutely distinct as the range of dormancy is continuous.
* Ratings based on overseas or commercial information only.
** Recommended for short term leys only.

Pest disease resistance ratings
HR Highly resistant — most plants will be unaffected by a severy attack.
R Resistant — light damage will occur under a severe attack but there will be little effect on the long term performance.
MR Moderate resistance — moderate damage will occur under severe attack but with satisfactory long term performance.
LR Low resistance — heavy damage will occur under severe attack which may affect long term performance.
S Susceptible — heavy damage will occur under moderate attack and long term performance will be affected.

To persist well, white clovers require a high annual rainfall of at least 700 mm (28") or irrigation. If however stands are killed by drought or bad management they are capable of regenerating from seed.

Given sufficient moisture they are capable of very high production on a wide range of fertile soils where they will persist well even under heavy grazing.

White clover should be sown into a fine well-prepared, weed-free seedbed at a rate of 1–2 kg/ha if it is the only clover in the mixture or at 1 kg/ha if other clovers are included.

It can be sown in either the autumn or the spring, but spring sowing has the advantage of avoiding the competition from annual weeds which tend to smother the clover seedlings. Spring sowing must be carried out early enough to allow the seedlings to develop a strong root system before the onset of hot summer conditions.

Because the seed size is very small, sowing depth sould be no more than 20 mm and if this cannot be achieved with the sowing equipment available, it is better to broadcast the seed and bury it with a light harrowing.

Grazing management should aim at preventing the grasses from becoming too tall, shading and suppressing the growth of the low-growing clover. Overgrazing, however, could cause the clover to become dominant to the extent where it will cause bloat in cattle.

The New Zealand or Grasslands Huia cultivar grows well during the spring, summer and autumn in the cooler districts.

Haifa has a higher winter production and is more drought tolerant than New Zealand. It is therefore suited to both irrigated and non-irrigated pastures and can be sown in districts where the annual rainfall is 650 mm (26") or more.

Tarma is very similar to Haifa in seasonal production and drought resistance and can be used in similar situations. Both Haifa and Tarma set seed well and are likely to spread to areas where they have not been sown in high rainfall districts. Their capacity to flower early and set a lot of seed, also enables them to regenerate particularly well following a drought.

Strawberry clover
Trifolium fragiferum

This perennial plant spreads by rooted runners producing good quality feed during spring summer and autumn providing moisture supplies are adequate. Its winter production is poor.

Strawberry will grow on a wide range of fertile soils in districts where the annual rainfall is at least 500 mm (20") or under irrigation.

It will withstand conditions that are too dry, too alkaline, or too saline for white clover and it will also tolerate poor drainage and waterlogging much better than white clover. These are the situations where strawberry clover should be used and will persist well under grazing.

Where white clover grows well it is more productive than strawberry and strawberry should not be used.

Strawberry clover should be managed in the same way as white clover. The main cultivar recommended is Palestine.

Red clover
Trifolium pratense

Red clover differs from white and strawberry clover as it does not regenerate from seed and it is shorter lived. It persists for only a few years.

It produces high quality feed in spring, summer and autumn but its winter growth is poor.

Red clover's use is restricted to short term pastures in irrigated situations on well drained fertile soils. In these types of pastures it will produce valuable early grazing and can be shut up to produce good quality hay.

It establishes easily and much quicker than either white or strawberry clover and should be sown at the rate of 5 kg/ha if it is the only clover in the mix, or at 2–3 kg/ha if other clovers are also being sown.

Sowing conditions should be the same as for white clover.

Because of its early vigour red clover should be grazed to prevent it from becoming dominant. It too can cause bloat in cattle and because some cultivars have a fairly high formononetin content they can also cause infertility in ewes if allowed to become dominant.

Redquin is a relatively new cultivar which has lower formononetin levels and apparently superior production capacity to the cultivar Grasslands Hamua which if often called New Zealand cowgrass in Australia.

Other cultivars include Grasslands Turoa and Grasslands Pawera. All of the Grasslands cultivars can cause infertility in sheep.

Pasture legumes

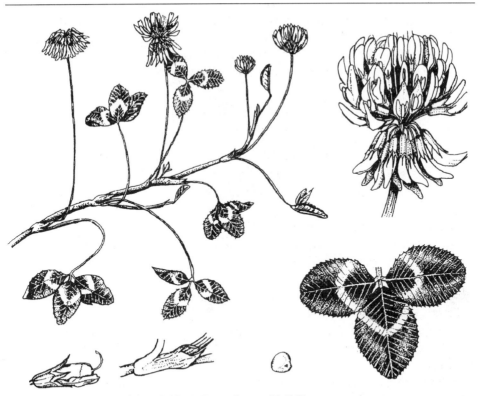

Figure 3.12 White clover *(Trifolium repens)*

Figure 3.13 Strawberry clover *(Trifolium fragiferum)*

Figure 3.14 Red clover *(Trifolium pratense)*

4

Sowing methods

Germination requirements

When sowing pasture seed in any situation the aim is to ensure maximum germination and rapid establishment.

This goal can be achieved only if the seed is correctly sown into a suitable environment.

For seed to germinate it requires sufficient moisture, a suitable temperature, air and in some instances a certain degree of light. If any of these requirements are missing it will not germinate.

Moisture

There must be sufficient moisture in the soil to enter the seed for germination. Seeding should coincide with suitable soil moisture conditions in autumn or spring.

Autumn is best for establishing non-irrigated pastures. This is the time when damp soil conditions are most likely to last long enough to enable satisfactory establishment in southern Australia. Earlier sowing increases the risk of "false breaks" to the season causing establishment failures. Prolonged dry periods may cause the seedbed to dry out too much to sustain the growth of seedling pastures. Later sowing increases the risk of waterlogging and low temperatures — inhibiting germination and establishment.

In the spring, sowing must maximise the time that damp soil conditions persist before the onset of hot dry summer conditions.

Spring sowing may start as soon as:

- ❀ soil conditions dry out enough to enable machinery onto the ground
- ❀ temperatures increase to permit germination to take place.

Temperature

Most seeds have an optimum temperature at which they germinate. This is usually between 25°C and 30°C and germination of many varieties will cease altogether at temperatures below 5°C or above 48°C.

Some seeds need to be subjected to periods of high and low temperatures before germinating.

In southern Australia, temperatures suitable for germination of most pasture varieties are most likely in the autumn and spring.

Air

Air is essential to provide oxygen for respiration in germinating seeds. Respiration converts insoluble starches in the seeds into soluble sugars which provide the energy for plant growth.

A well aerated, well drained seedbed will ensure that conditions such as waterlogging do not create a lack of oxygen and inhibit germination.

Light

Different seeds have different needs for light. Some, particularly grasses, will not germinate at all if planted too deep for light to penetrate, while others, including many legumes, will germinate as well in dark conditions as in light. Seeding depth is therefore critical. Seed planted too deep may not germinate.

Another critical effect of seeding depth is that seed placed too deep in the soil may not have sufficient stored energy reserves to allow the shoot to reach the soil surface. Because pasture seeds are small, they should be placed close to or on top of the soil surface.

Seedbed preparation

Planning is essential to the success of any pasture establishment or renovation program. Not only is it necessary to create conditions that will maximise germination rates but conditions must also ensure that the resulting seedling pasture establishes satisfactorily.

To do this the seedbed must be prepared to:
- ❀ control weeds
- ❀ conserve moisture
- ❀ correct nutrient deficiencies
- ❀ to modify the soil structure,

so that the pasture seed can be sown into a well aerated, well drained, firm fine seedbed without a cloddy surface. Such a seedbed will enable the seed to be sown evenly at the desired depth or 1–2 cm (shallower in heavy soils, deeper in sandy soils) in soil conditioned to encourage root and shoot growth.

A rough seedbed will result in poor seed germination and poor early growth favouring weed rather than pasture growth. The aim of any seedbed preparation technique is to prepare an environment that provides good soil to seed contact and allows the seed to be placed at a constant depth. The traditional method of cultivating the soil to prepare the seedbed can achieve this goal and results in higher success rates than other methods of pasture establishment.

In the past, direct drilling could not achieve the same success in pasture establishment, because sowing depth, soil to seed contact and weed control were not optimal. Advances in direct drilling equipment have improved the problems of soil to seed contact and sowing depth to the point that this method of pasture establishment can be as successful as traditional cultivation techniques. Direct drilling also relies heavily on weed reduction strategies using livestock and herbicides.

Weed control

Weeds compete with pastures for moisture, nutrients and light. Newly sown pastures, particularly less vigorous perennial varieties, are very susceptible to competition from weeds and can easily be smothered and choked out before they have had a chance to establish.

Weeds must be controlled if establishment or renovation programs are to be successful.

Each area to be sown must be inspected, preferably during the previous growing seasons and their weeds identified and assessed. Effective control measures must then be planned and put into action. Control may take years or only a few weeks depending upon the weeds and the severity of the infestation.

Serious weeds may need to be controlled through heavy grazing and with the use of herbicides. The area may also need to be cropped for one or two years to enable control with a combination of cultivation and selective herbicides applied to the weeds in the cleaning crop.

Less serious annual weeds may be controlled relatively easily and quickly using cultivation and/or selective herbicides.

Weed control will be easier if weeds are prevented from seeding in the years before sowing pastures. This can be done by using herbicides, mowing or heavy grazing.

Cleaning crops

These are crops that enable selective herbicides to be used on weeds without damaging the crops. They are often used as a means of preparing land to sow pastures.

Oats or wheat are generally used as cleaning crops for the control of broadleaf weeds. Oats are more popular in high rainfall districts. They provide useful winter grazing and can also be cut for hay or reaped.

Grass weeds are controlled in cleaning crops such as lupins, peas or faba beans.

In the spring and summer fodder crops may be used as cleaning crops.

Burning

Burning is usually used only when preparing a seedbed if there is too much trash on the surface to be turned in with cultivation equipment. However, besides making the job of cultivation easier, burning helps to control weeds by destroying weed seeds on the surface of the soil and by breaking down hardseededness.

This makes later weed control measures with cultivation or herbicides more effective.

Ploughing

Ploughing is the first step in "working up" a seedbed for the traditional method of pasture establishment. It should be done early enough to:

- enable the organic matter turned into the soil to decompose
- allow time to kill subsequent germinations of weeds
- prepare a fine clean seedbed.

Depth of ploughing should be adjusted to ensure that weeds are uprooted and turned in but avoiding unneccessary disturbance of any layers of clay or stone below the topsoil.

On sandy soils blade ploughs, that leave the trash on the surface, will help to prevent erosion.

Cultivation

Cultivations, usually two, should be carried out as seasonal conditions dictate, working the ground when it is damp to kill successive weed germinations and to create a fine seedbed. The ground should be cultivated to about half the depth of the initial ploughing.

Rolling and harrowing

Rolling before seeding helps to create a better seedbed on light and cloddy soils. In spring it is particularly useful for breaking down clods. Rolling compacts the soil leaving fewer large air spaces so that the seed and soil come into close contact and helps to ensure that soil moisture is held near the surface, close to the seed, by increasing capillary action that would not be as strong in less tightly packed soils.

Rolling before seeding also helps to level the seedbed and to bury stones brought to the surface by cultivation. This enables better and more even depth control with seeding equipment.

Rolling after seeding will further compact the seedbed, but extra care should be taken on fine sandy soils to avoid the risk of sandblast damaging emerging seedling plants, and on soils that form a hard crust when wet as the crust is likely to prevent seedlings emerging.

Light "pasture" harrows can be used after cultivation to control weeds, break up clods and to level the seedbed. Trailed behind seeding implements they ensure that the seed is buried and covered thoroughly.

Cover crops

Cover crops are crops sown to protect establishing pastures.

On light sandy soils seedling pasture plants such as lucerne can be protected from sandblast damage by firstly seeding a cereal such as cereal rye, at a low rate (5–10 kg/ha), so as not to provide too much competition to the pasture. The pasture seed should then be sown as the cover crop starts to emerge. By the time the pasture emerges the cover crop will be tall enough to prevent sandblast damaging the pasture.

In high rainfall districts oats are often used as a cover crop for perennial pasture mixtures to ensure some grazing from the sown area while the pasture is establishing. The oats are usually sown with the pasture mixture at about one third of the rate at which it would be sown alone, that is in high rainfall areas at about 40 kg/ha. Care needs to be taken to ensure that the oats are not allowed to smother the pasture varieties, and that weeds, insect pests and grazing livestock do not cause too much damage to the pasture, defeating the purpose of the cover crop.

Chou moellier is another spring cover crop often used to protect perennial grasses, particularly phalaris. This tall brassica will enable the grasses to establish while it provides useful summer and autumn grazing. The grasses can be oversown with annual legumes in the autumn.

Aerating

Aerating is another method of cultivating that is used to help renovate existing pastures on hard setting compacted soils. It involves deep ripping the ground to a depth of about 40 cm when the soil is dry. This aerates the soil, improves water penetration and retention and encourages deeper root growth of pasture plants, making them more drought resistant and possibly extending their growth period.

There is a whole range of equipment suitable for aerating the soil. These include the "agro-plow", the "Wallace" and the "Yeomans" ploughs and other similar machines. All have high running costs due to the high tractor horsepower required, tractor tyre wear and point wear on the implement.

Aerating should be carried out on the contour to maximise water retention.

Fertilising

This topic is dealt with in greater detail in Chapter 7 but also warrants mentioning in relation to seeding.

The correct use of fertilisers increases establishment vigour and subsequent production and it helps to prevent establishment failures.

Soils should be tested to determine their requirements for the major elements — nitrogen, phosphorus and potassium (N, P, K) and to find trace element deficiencies. Trace element deficiencies of copper (Cu), zinc (Zn), molybdenum (Mo) and cobalt (Co) can be common in different soils throughout southern Australia. All need to be corrected before seeding.

On very acid or alkaline soils extremes of pH may cause some plant nutrients to be unavailable or to reach toxic levels. Aluminium is an exception as it is not a plant nutrient, but reaches toxic levels in some soils with a low pH. Lime should be applied to very acid soils to overcome this situation, but to be most effective it should be incorporated into the soil.

If seed has to be mixed with fertiliser for sowing, make sure that the time that the seed is actually mixed in with the fertiliser is kept to a minimum to prevent damaging the germination.

Sowing methods

Drilling into a prepared seedbed

Drilling is certainly the most accurate method of distributing seed at a required rate. If seedbeds are suitable and seed is sown at the correct depth then drilling will also produce the best germination.

In the past it was common to use a small seed box attached to a cereal combine for pasture establishment into a prepared seed bed. Today this is still a common method of pasture establishment, but there are a variety of new types of pasture seeding equipment available. Modern types of pasture seeding equipment can be more accurately calibrated to the correct sowing rate and can more accurately place the pasture seed at the correct depth. Other advanced features include disc and tyne design for direct drilling and sod seeding, placement of fertiliser in a band with the seed or in a band below the seed and coulters to create less soil disturbance.

For the best results it is necessary to follow these rules when drilling seed.

    ❀ Always calibrate the sowing rate of the drill before starting seeding. All pasture seeds and pasture seed mixtures vary in size, density, shape

and texture. All affect the rate of flow through a drill. Calibrate as follows:
1. Measure the effective width of the drill — the distance between the outside discs or tynes.
2.. Measure the circumference of the wheels of the drill in metres.
3. Jack the drill up so that the wheels may be turned by hand.
4. Half fill the seedbox with the seed to be sown and with buckets placed under the seed delivery hoses turn the wheels to check that all outlets are free and working.
5. Empty the buckets back into the seedbox and replace them under the delivery hoses.
6. Set the drill cogs to give you the desired sowing rate according to the manufacturer's instructions.
7. Turn the wheels ten full revolutions.
8. Weigh the seed collected in the buckets.
9. Calculate:

$$\text{Sowing rate in kg/ha} = \frac{\text{Wt of seed collected} \times 10\ 000}{\text{effective drill width} \times \text{wheel circumference} \times 10}$$

10. Adjust the cogs accordingly if the rate does not agree with the rate you require and retest until the rate is correct.

Fertiliser rates should be checked in the same way.

- Try to keep the seedbox half full for as long as possible when working. The seed will run faster from a full box than an empty one and the seeding rate will decrease as the box empties.
- If working in damp or wet weather check the delivery hoses regularly to ensure that they do not become blocked.
- The discs or tynes should be set to sow the seed at a constant depth of 0.5 to 1 cm.
- Drive slowly to ensure that discs or tynes penetrate evenly and that the seed is covered.
- If depth control is too difficult to control remove the seed delivery hoses from the shoes on the tynes or discs to drop seed on the surface of the soil. Bury it by trailing light pasture harrows or a chain.

Providing these steps are followed an even establishment of pasture will result.

Sod seeding

Sod seeding involves direct drilling seed into an undisturbed seedbed. It is an excellent method for renovating run down pastures and sowing new improved pasture varieties on non-arable land. Sod seeding should be carried out in conjunction with "chemical ploughing" to control weeds.

This method of sowing pasture seed has many advantages over sowing into a prepared seedbed or broadcasting.

- ❀ Preservation of existing pasture varieties in the sward — particularly the slow to establish perennial grasses.
- ❀ Quicker sowing — it enables large areas to be sown quickly and the land is not out of production for as long as it would be if it was cultivated to prepare a seedbed. This gives increased flexibility to the overall farm management program and reduces grazing pressure on the paddocks that have not been sown.
  The speed of the operation also makes this method of seeding ideal for overcoming periods of feed shortage when rapid establishment is needed after bushfires, droughts and other disasters, or to restore the balance in clover dominant pastures.
- ❀ Reduced risk of weed infestation — Weed seed on the surface has little chance of being buried and the competition from the existing pasture plants helps to prevent weeds from becoming established.
- ❀ Easier sowing on non-arable land — sod seeder discs or tynes ride over stumps and stones.
- ❀ Less surface stone — loose stones are not brought to the soil surface by sod seeding to harm machinery or livestock.
- ❀ Wetter sowing — the undisturbed seedbed enables seeding to be carried out when conditions are too wet to enable seeding into a prepared seedbed.
- ❀ Less pugging — damage caused by livestock pugging up the ground when grazing under wet conditions is reduced because of the compact seedbed.
- ❀ Reduced erosion — the risk of erosion from both wind and water is reduced because the ground remains undisturbed and to some degree covered.
- ❀ Preservation of existing plants — new species can be added to a pasture with a minimum amount of damage to the existing pasture plants.
- ❀ More grazing — loss of grazing time is minimised.
- ❀ Correct fertiliser placement — fertiliser is placed close to the seed to boost rapid establishment and growth.
- ❀ Soil improvement — on heavy soils sod seeding assists in aerating the seedbed and aids water penetration and retention.

Disadvantages of sod seeding must also be taken into account when you are considering using this sowing method. They include the following:

- ❀ Lower germination percentage — the percentage of seed that germinates and establishes is not as high as that of seed sown into a prepared seedbed.
- ❀ Competition from existing plants — competition from pasture plants in the existing sward hinders establishment of the sod seeded varieties and may cause seedling mortality,
- ❀ Pests — the existing sward may harbour pests such as Redlegged earthmite and lucerne flea that are likely to cause severe damage to emerging seedlings if not controlled.

Sowing methods

- ✽ Reduced grazing during establishment — the existing sward must be grazed sparingly while the sod seeded plants are establishing.

There are two types of sod seeders. Some have discs to open up the furrows in the sward into which the seed and fertiliser are dropped. Others have narrow pointed tynes to create furrows.

Tyned sod seeders are best for sowing rough country, undulating or stony ground and hard setting soils. The "Baker" boot or a similar concept in the "T" boot are one the major advances in foot design for sod seeding equipment. The shape of the foot creates an inverted T-slot in the soil which has minimal surface disturbance but shatters the soil around the seed to create an ideal environment for the seed to germinate without drying out. The "Baker" boot is designed to band fertiliser below the seed, to minimise fertiliser burn, while the T-boot places seed and fertiliser together.

Subsoil ploughs can also be fitted with seeding boxes, the simplest being a box that distributes seed through the vibration of the plough. With these types of seeders, pasture seed is placed on the rip lines produced by the subsoiler.

Disc sod seeders give better results on lighter soils that are penetrated more easily. The triple disc seeder is the most sophisticated disc pasture

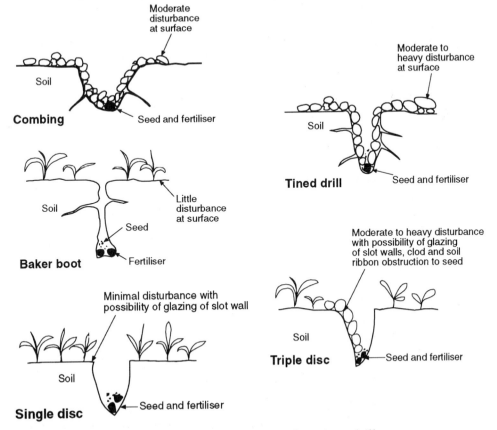

Figure 4.1 Seed placement with various drills

seeding equipment available. Unlike conventional disc seeders which tend to leave pasture seed uncovered, the triple disc creates minimal disturbance and provides good seed to soil contact. Triple disc seeders do not work well in wet soil conditions.

For the best results when sod seeding adhere to the following rules:

1. Prepare the area to be sod seeded by removing surface debris such as sticks, stumps and stones and graze the area hard, preferably with sheep, in the late summer/early autumn, to bare as much ground as possible and assist even penetration by the tynes or discs of the sod seeder. Do the same during late winter for sod seeding in spring.
2. Prevent weeds from seeding in the previous year by using desiccant herbicides, mowing, or hard grazing. Then, following the opening rains, chemical plough if necessary.

    It may also be necessary to commence a pest reduction program in the year prior to sowing. This can include spraying for redlegged earthmites in the spring before sowing to reduce their numbers when the pasture is sowed the following year.
3. Sod seed into a damp seedbed as early as possible in the autumn, when there is little risk of the seedbed drying out and before the ground gets too cold, or in the early spring to ensure that the seedbed remains damp for as long as possible.

    If the soil is too wet it is difficult to cover the seed effectively. If the soil is too dry then germination will be poor and the furrows may collapse over the seed burying it too deeply.
4. Make sure that the sod seeder is operating properly, that the sowing depth is correct and that the seed and fertiliser tubes do not block.
5. Whenever possible sow varieties suited to the area that have plenty of seedling vigour.
6. Compensate for anticipated poor germination and establishment, (compared to sowing into a prepared seedbed) by increasing the "prepared seedbed" seeding rate by about 50 per cent.
7. Operate the sod seeder no faster than 5 km/hr. Higher speeds will inhibit penetration and leave the seed unevenly covered resulting in poor establishment.
8. Do not skimp on seed or fertiliser and inoculate and lime pellet legume seed.
9. Graze sparingly while the new sown pasture varieties establish. Established varieties must not be allowed to smother the newly sown plants. Ongoing invertebrate pest management is also necessary.

Broadcasting

Broadcasting is the oldest method of sowing seed known to man. Today seed, mixed with fertilisers, is broadcast from agricultural aircraft, ground operated fertiliser spreaders and by hand in a variety of different situations.

Agricultural aircraft sow seed over land that is too rough, steep, or wet to spread from the ground.

Ground operated fertiliser spreaders are used to seed areas that cannot be reached or sown using conventional seeding equipment in partially cleared country, on rocky ground and amongst trees, and also to oversow existing pastures when applying fertiliser.

Broadcasting is the cheapest method of sowing seed, but it is the least reliable and establishment failures are not uncommon. However, if the following rules are followed you will greatly reduce the risk of failure.

1. Prepare the ground to be broadcast as you would for sod seeding by cleaning up the surface debris and grazing it hard during late summer and early autumn to bare as much ground as possible.
2. Control the weeds as you would if you were sod seeding.
3. Broadcast when the ground is damp and likely to stay damp, in the autumn or spring.
4. Attempt to cover the seed if equipment is available by harrowing, raking etc. or alternatively by heavily stocking the area with sheep for two or three days to pug the seed into the damp seedbed.
5. Seed and fertiliser should be mixed evenly just prior to spreading and the time that the seed is left mixed in the fertiliser kept to an absolute minimum. Acid based fertilisers such as superphosphate can destroy the viability of seed left mixed in them so avoid delays once the seed and fertiliser are mixed.
6. Inoculate and lime pellet legume seeds.
7. Compensate for anticipated poor germination and establishment by increasing the seeding rate to about double that you would use if drilling into a prepared seedbed.
8. To ensure even distribution of the seed reduce the application rate to half of the desired rate and spread the entire area twice as illustrated.

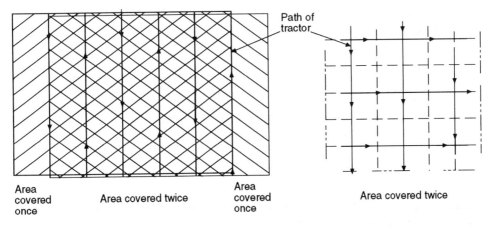

Figure 4.2 Broadcastng patterns, for double coverage and even distribution

9. Grazing should be carefully managed until the pasture has had sufficient time to establish properly.
10. Insect pests should be controlled if necessary. Seed should be treated to reduce loss by ants.

Chemical ploughing

This process is often used in association with sod seeding or broadcasting in a renovation program to control annual weeds.

Desiccant herbicides such as diquat (Reglone®) are used to control broadleaf weeds, paraquat (Gramoxone®) to control grasses or a mixture of both (Sprayseed®) to control both broadleaf weeds and grasses. Systemic herbicides such as glyphosate (Roundup®) can also be used to control a broad range of weeds through techniques such as "spray grazing" and "spray topping".

To ensure good results:
- Graze the area to be treated bare during summer.
- Following opening rains graze paddock hard with sheep until two or three days before spraying.
- Spray four weeks after the opening rains.
- Ensure the weeds are fresh and growing vigorously at the time of spraying.
- Avoid spraying if the plants are stressed, covered in dust or dew.
- Mix the herbicide with clear water, free from clay or silt.
- Apply the herbicide accurately at the recommended rate with a boom spray.
- Sod seed as soon as possible, within five days of spraying.
- Difficult to kill weeds (e.g. dock, sorrel and salvation Jane) should be treated with a systemic herbicide such as glyphosate.

Summary of seeding requirements

Successful seeding programs depend upon the following:
- Good planning
- Thorough seedbed preparation.
- Selection of suitable pasture varieties
- Use of certified, weed free seed
- Adequate rates of seed (sow a smaller area rather than skimp on seed)
- Inoculation of legumes
- Correction of extremes in pH
- Adequate fertiliser rates, including trace elements
- Correct sowing depth and even distribution of seed and fertiliser
- Timing to coincide with suitable weather/soil conditions for germination and establishment
- Follow up management of grazing, weed and pest control.

5

Invertebrate pest control in pastures

Invertebrate pests can be controlled in the following ways:
- use of pesticides
- biological controls
- resistant varieties.

The use of all three methods is called integrated pest management.

Pesticides

Pesticides are generally the quickest form of control when a pest and subsequent damage has been diagnosed. It may also be the most costly, and the cost-benefit of spraying pastures needs to be seriously considered in some circumstances, especially where pastures are likely to be re-invaded after spraying.

Safety precautions

All pesticides should be regarded as poisonous and potetially dangerous. Modern technology is continually developing new pesticides and modifications to existing ones, therefore it is extremely important to READ THE LABEL every time, even if the operator has used the same type of insecticide before. Registration laws require that the label also contains safety precautions. Check the current registration of pesticides stored on the farm.

Invertebrate pests are living organisms with many internal organs similar to those of humans. The method of action of pesticides is to affect certain organs within the pest. Therefore the compounds which affect pests can also affect humans and livestock.

Pasture management

There are three main ways in which poisoning can occur from pesticides: through ingestion, inhalation, and absorption through the skin. Therefore safety recommendations are designed to minimise the entry into the body by any of these methods.

It is important to reduce the exposure of livestock to pesticides as much as possible. Stock should be moved well away from the area to be sprayed, and it is the responsibility of the operator to be aware of spray drift. Withholding periods are a legal requirement to prevent contamination of produce that could jeopadise public health and markets.

Preparation

- Select the safest and most appropriate pesticide. (Make sure you have the pest you think you have!)
- Use only registered chemicals in your state. Registration may vary between states.
- Read the whole label, including withholding periods for grazing animals.
- Use only according to the label. It is illegal to do otherwise.
- Have appropriate first aid equipment accessible.
- Expose the minimum possible skin area, using chemical resistant clothing, boots, hat and gloves.
- Wear a respirator and follow all safety precautions.
- Check the spray equipment for correct rates, and calibrate if necessary.
- Calculate and prepare the correct amount of pesticide for the area to be sprayed. Avoid having to dispose of unused pesticide.
- Handle concentrates with care — remember how dangerous even the final mixture is!
- Use a proper measuring container.
- Wash all containers used in preparation, without contaminating water supplies.
- Dispose of empty pesticide containers properly and safely.
- THEN wash yourself using soap, before eating or smoking.

Spraying

The most common machines are either the boom spray or mister. The mister relies on drift to apply the pesticide (usually an insecticide), and it is important that the operator knows exactly where ALL the mist is going. These machines should not be used in extremely windy conditions, because the application rate will not be correct, and the mist will probably drift to other areas. This also applies to "backyard" knapsack sprayers or similar. The following points should be remembered when spraying:

- Make sure all jets are working, using water only.
- Wash down the machine after concentrate has been added.

- Spray across the wind if possible. This will ensure the operator is always exposed to fresh uncontaminated air.
- Do not go back over the same area. This is sometimes tempting, especially if you have mixed too much. Stock withholding recommendations are made on certain rates, and in some circumstances beneficial organisms such as parasites may be killed with higher rates.
- Wash out spray tank upon completion — it is easy to forget what was in the tank, and some pesticides are not compatible.
- THEN wash yourself using soap, before eating or smoking.
- Wash all clothing on its own.

Rates

All information regarding correct rates, timing of applications and stock withholding periods has to be displayed on the label BY LAW. However, it is important that the operator has the necessary literature on hand to select the correct pesticide before purchase.

Consult the your Department of Agriculture literature for information about application rates, calibrating spray equipment, and safety.

Users of pesticides are advised to undertake the National Farm Chemical User's Course. Such training will bring them up to date with all aspects of chemical usage. These courses of about two days' duration are available throughout Australia.

Timing of treatments

Pest populations fluctuate depending on environmental influences. To properly manage pastures, it is imperative that an understanding of the most common pasture pests for an area is gained and that pests are controlled before significant damage to the pasture occurs. Department of Agriculture fact sheets are available for the most important pasture pests and should be referred to whenever possible.

Other control methods

Biological control is the use of other organisms to control a pest. In the case of another insect, it is either a parasite or a predator. By definition, a parasite needs the pests for its own survival, it cannot live on its own. A predator, on the other hand, preys on a range of organisms and does not need a specific organism in order to survive. There are also many diseases including viruses and fungi that will proliferate on pests.

The discovery and implementation of a biological control agent is usually a long term operation, because it has to be ascertained that this agent will not become a pest in its own right. This operation usually involves considerable research and quarantine procedures. Most biological control agents are introduced from other countries, where the pest in question originated from and may or may not be a problem.

Resistant varieties play a major role in overcoming damage by pests. Plant breeders use genetic techniques to incorporate resistance into plants that are already adapted to the environment. If a plant is resistant, it will not be attacked by a pest or will suffer less damage. To breed a resistant variety can take up to ten years or more, depending on the pest and the situation. It is possible to breed resistance to more than one pest into any given plant variety.

Integrated control uses a combination of pesticides, biological control, resistant varieties and cultural practices to control insect pests. Each area of integrated control needs to be compatible with the other. For example, it is important that the pesticides used in an integrated control program are not effective on other parasites and predators.

Invertebrate pests

The following are major pasture pests throughout Australia, their effects and recommended controls. Note that specific chemical control measures given may change due to developments in products and pesticide resistance. Consult your local Department of Agriculture or agronomist for recommended pesticides.

Barley grub (or southern armyworm)
Persectania ewingii

Host range and distribution
Barley grub, or southern armyworm, is found in the temperate zones of southern Australia. It is a destructive insect pest of barley crops and will also attack wheat and oat crops. It can cause serious losses in phalaris, cocksfoot, fescue and ryegrass seed crops.

Description
The barley grub is the larva (caterpillar) of a native night-flying moth, and grubs grow to about 4 cm in length. The larvae are smooth and are marked by longitudinal stripes running the full length of the body. Body colour is variable depending on the host but the head, "collar" behind the head, and the tail piece are always brown with three longitudinal white lines.

The larvae transform into dark brown pupae near the soil surface. The moth has greyish-coloured forewings, streaked with brown or black. Near the middle of each forewing there is a white "fish-shaped" mark and the rear end of the wing contains from five to seven crossed white lines.

The eggs are small white spheres, laid in batches of 20–100 around the sheathed base of grass leaves. In phalaris, eggs are laid in the sheath surrounding the emerging seed head.

Life cycle and biology

There are generally two generations of barley grub each year — the winter generation and the spring generation. In southern Australia the spring generation is of the most economic importance.

Moths of the autumn flight lay eggs on grasses, which produce the winter generation. They feed on different grasses but mainly barley grass, and young cereal crops. They grow slowly due to low temperatures, but by the end of August they are fully grown and pupate.

After about a fortnight, depending on temperatures, moths begin to emerge and continue to do so over a period of about six weeks. They then lay eggs in grasses, cereal and grass seed crops, which hatch into the spring generation of larvae which occurs from September to December.

With the warmer spring temperatures, the larvae grow much more rapidly than the previous generation, completing larval development in about three weeks. In the higher rainfall areas the oldest larvae of the spring generation can undergo a further generation during December.

Plant damage and symptoms

In phalaris and cocksfoot seed crops, young larvae damage heads emerging from the sheath. Older grubs may also feed on the heads. White areas on the sheath indicate barley grub damage.

In ryegrass young larvae feed on the flowers and developing seeds. High populations can severely reduce seed production. In cocksfoot seed crops, serious damage also occurs from larvae eating stalks at the nodes.

In barley the most serious damage occurs as the crop ripens. Larvae attack the green tissue just under the head, which is the last part of the barley plant to ripen. As a result the untouched heads are severed from the plant.

One larvae per square metre can cause serious damage and from mid-October onwards a close watch must be kept for infestations. One large

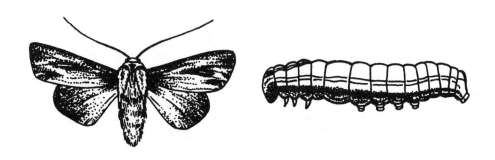

Figure 5.1 Barley grub (*Persectania ewingii*)

grub can cut off up to seven heads of barley a day. Greenish yellow pellets of excreta on the ground demonstrate a relatively high larvae population. An insect sweep net will also help in assessing grub numbers in a crop.

In oats damage occurs in much the same way as in barley, although subsequent grain loss is due to a less compact seed head, while damage in wheat is concentrated on the head itself. This is because the heads of wheat are the last part of the plant to mature.

In pastures damage may occur until as late as the end of December. Infestation by invasion from other paddocks may also occur. Occasionally, as heavily infested grasslands dry up or are eaten out, the larvae migrate to surrounding areas, hence the common name armyworm.

Control

The timing of spraying insecticides is important, and although spraying in cereals and phalaris can be delayed until heads emerge, it is important to spray ryegrass and strawberry clover seed crops when larvae numbers reach one to two per square metre. Delaying as long as possible will allow control of larvae that move in from other crops.

Recommended insecticides are Chlorpyrifos (Lorsban®) and Fenvalerate (Sumicidin®) in cereals, Chlorfenvinphos (Birlane 50®) and Chlorpyrifos (Lorsban®) for pastures and Chlorpyrifos (Lorsban®) for phalaris, strawberry clover and ryegrass seed crops. Check current recommendations for rates and alterations and government registration of chemicals.

There are no biological agents or resistant varieties available.

Blue-green aphid
Acyrthosiphon kondoi

Host range and distribution

Blue-green aphid, a native of Korea and Japan, was first discovered in eastern Australia in June 1977. Blue-green aphid has become a major pest of annual medic pastures throughout southern Australia. It has also become a pest of lucerne.

Description

Blue-green aphid closely resembles other commonly found aphids, although it has a unique blue-green colouration and waxy appearance. It is approximately 3 mm long and has two prominent spines extending from the rear of its back.

Life cycle and biology

Blue-green aphids are found only as asexually reproducing females in Australia. Optimum multiplication occurs at temperatures of 10–18°C with 30–40 per cent relative humidity. At least 20 generations may occur in a year. As populations build to large numbers the last produced aphids are nearly all winged forms.

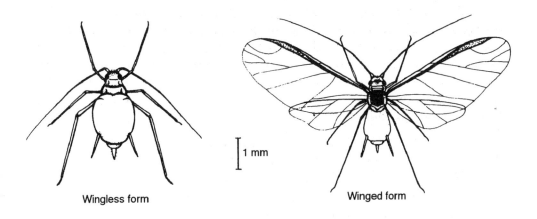

Figure 5.2 Blue-green aphid (*Acyrthosiphon kondoi*)

Plant damage and symptoms

Early blue-green aphid damage is detected by the curling under of leaves. An easy check is to shake a few plants on to the hand or a piece of paper.

Blue-green aphid tends to stunt plant growth, and will not kill the plant directly. The winged form flies from the plant before serious symptoms of attack appear, leaving the plants alive but in a seriously weakened condition. This then leaves the plant subject to attack by several virus and fungus pathogens that reduce plant growth even further. The aphid is said to be acting as a vector in this situation.

Control

Pirimicarb (Pirimor®) and Demeton-S-methyl (Metasystox®) are recommended for use in annual medics and lucerne. Check current recommendations for rates, as rates which are too high may kill important predators and the spotted alfalfa aphid parasites. Check current registration of chemicals.

At present there is no parasite suitable for blue-green aphid control, although the common ladybird is a predator.

Considerable research has been undertaken to develop blue-green aphid resistant strains in annual medics and lucerne. Check current medic varieties for resistance to this pest.

Cabbage centre grub
Hellula hydralis

Host range and distribution

Cabbage centre grub is predominantly a pest of cruciferous crops such as chou moellier, rape and turnips but will also attack cruciferous weeds such as wild turnip, ward's weed and mustard.

Description

The fully grown larvae is about 15 mm long and is greenish in colour with several longitudinal reddish brown stripes along the abdomen. The adult is a small pale coloured moth about 15 mm long. The forewings are a pale brown colour except for the hind margin which is slightly paler. A very small "eyespot" in the hindwings also helps in identification. When the moth is at rest the wings are held horizontally. The moth lays eggs on the host plant, which hatch in about 7–14 days.

Life cycle and biology

Upon hatching young larvae bore into the leaf and feed on the tissues between the upper and lower leaf cuticles, which is typical of leaf miner attack. The larvae then grow through several instar stages (moults) until they are too large to feed between the leaf cuticles. They then emerge, usually on the lower surface of the leaf, shelter under a "silken tent" and feed on the leaf. The larval stage lasts for about four weeks and then the caterpillars crawl down into surface litter or into the soil and pupate. Pupation lasts for about one week, upon which time the next generation of adults emerge.

Several generations occur each year. Moths of the first generation usually occur in the spring, and are the result of overwintering pupae. This generation grows almost exclusively on cruciferous weeds including wild turnip and mustard and usually requires about two months for completion due to the cooler conditions.

Subsequent generations during the summer months take place within about five to six weeks, and although these generations may occur on any crucifers, the most economic damage occurs on summer fodder crops such as rape, turnips and chou moellier, causing fodder loss. In more severe cases, total crop loss can occur.

Plant damage and symptoms

As well as attack of the leaf cuticles, older larvae feed on the stems and fleshy petioles, always under a silken shelter. The presence of these larvae under webbing, together with adhering excreta is characteristic of the cabbage centre grub.

Control

Chou moellier, the most important of the cruciferous fodder crops, is grown primarily for late summer and early autumn feed in high rainfall areas. Damage by cabbage centre grub can be minimised if plants are established early in the season.

Due to larvae being sheltered by the silken webbing, control by insecticides is not always successful. Check current recommendations for available insecticides.

There are no parasites of cabbage centre grub available, nor predators of any economic significance.

There are also no resistant varieties available.

Climbing cutworm (or native budworm)
Helicoverpa punctigera

Host range and distribution

The climbing cutworm is a native insect of Australia which has become a pest, particularly of leguminous crops, feeding mainly on the buds, flowers and fruit parts of plants. It will also feed on plant foliage. It is a major pest of field peas, lucerne, lupins, sub clover and annual medic seed crops. Linseed, rapeseed and sunflower crops, as well as maize and tomatoes are attacked, and occasionally serious damage occurs in cereal crops, and stone and pome fruits to a lesser extent.

Description

The adult is a pale cream or buff coloured moth, females being slightly darker in colour. The hindwings are whitish with prominent dark patches on the rear margins. The wing span is approximately 30–45 mm. Eggs are dome-shaped, flattened at the base and are about 0.5 mm in diameter. When first laid they are white to greenish yellow, but they turn brown just before hatching. Larvae (caterpillars) grow to 50 mm in length and vary in colour, depending on their host. Generally body colour can be green, yellow, pink or reddish brown to almost black. *Helicoverpa* can, however, be distinguished by a broad yellowish stripe along each side of the body. Unlike other caterpillars, the rear part of the body is angled sharply downwards.

Life cycle and biology

There are usually two or three generations each year. Moths begin flying in late August to early September, and must feed on nectar before they can lay. They live for about two to four weeks. Each female can lay up to 1000 eggs. Warm weather in late winter and early spring favours egg laying. Eggs are laid singly on crops and weeds and take up to 12 days to hatch, depending on temperature. This first generation is referred to as the "spring brood" and is of particular concern in peas and lupins. After feeding for about four weeks, mature larvae leave the plant and pupate in the soil.

Adults generally begin to emerge after approximately two weeks, depending again on temperatures. Larvae that hatch from this second generation are referred to as the "summer brood", and are particularly a problem in lucerne seed crops. A further generation may occur in the autumn.

Plant damage and symptoms

In leguminous and oilseed crops, larvae usually feed on plant foliage before burrowing into the developing pods. Young larvae have a strong preference for buds and flowers, and with heavy infestations the ovaries may be damaged, therefore reducing yields considerably. Monitoring of egglaying and numbers of eggs present is of assistance in anticipating high infestations. Crops should be checked to enable adequate time for control measures to be implemented.

Control
The use of insecticides has proven to be the most suitable form of control, but timing is of major importance. Moth flights have to be monitored, preferably in the late afternoon and evening when females move from plant to plant to lay eggs. Small larvae are much easier to control, although it is best to wait until a majority of the larvae have hatched. If one or two larvae can be found per square metre, it is time to spray.

Fenvalerate (Sumicidin®) is recommended for control of *Helicoverpa* sp., although check current recommendations for alternatives and rates. Check for current registration of chemicals.

There are no biological controls or resistant pasture varieties available.

Black field cricket
Teleogryllus commodus

Host range and distribution
The field cricket is a native insect which occurs particularly on heavy black cracking soils. In some years it reaches plague numbers causing extensive pasture damage. The insect attacks the seeds of pasture grasses and legumes.

Description
The female field cricket is a shiny black insect about 30 mm long and carries a long slender ovipositor (for egglaying). The male has more wrinkled forewings and produces a trilling call by vibrating its wings. Eggs are pale yellow to white, sausage shaped and about 3 mm long. Nymphs, which are like wingless miniatures of the adult, hatch from the eggs. The earliest nymphal stages are ant-like in appearance, but as they grow through a series of instars (moults) they begin to take on the appearance of the adult.

Life cycle and biology
There is only one generation of field crickets each year. Female adults lay large numbers of eggs in the soil after rain in late summer and early autumn, and then die within a month. The eggs remain dormant during the winter and hatch about mid-November. Nymphs are rarely seen as they hide in cracks in the soil during the day, and come out to feed at night. Adults begin to appear during late February, and from then on infestation occurs.

The development of more productive pastures on soils subject to cricket attack has favoured an increase in the cricket populations by providing more feed and ground cover, which reduces egg mortality through drying out. The development of plagues is favoured by wet conditions during winter, continuing up to egg hatching in mid-November.

Plant damage and symptoms
Grasses and clovers are eaten out or plant stalks are cut off at the base and dry out quickly. Both dry and green foliage is attacked. Strawberry clover

plants may be selectively eaten and the roots attacked which causes high plant mortality and poorer pasture production in the following season. Sub clover seed will also be attacked and eaten, along with baled hay left in the paddock for autumn grazing.

Control

Chemical spraying and grain baiting with insecticides are the two alternative methods of field cricket control. Spraying has proven to be more effective, however pasture cover must be low to allow adequate insecticide penetration. Grazing or hay-cutting by late December is practised in areas prone to field cricket damage. It is important to spray when cricket movement and subsequent feeding is imminent, generally after summer and autumn rains.

Technical Maldison (Malathion ULV®) is recommended for spraying and Maldison (Malathion®) mixed with wheat for baiting. Refer to current recommendations for alternatives, rates and current registration of chemicals.

There are no biological controls or resistant strains available.

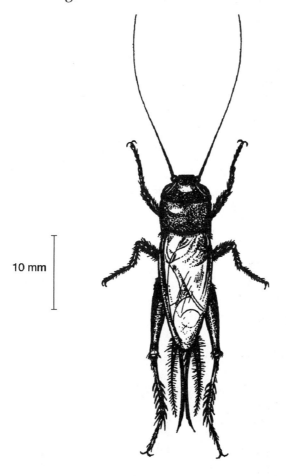

Figure 5.3 Field cricket (*Teleogryllus commodus*)

Lucerne flea
Sminthurus viridis

Host range and distribution

Lucerne flea, a native of the Mediterranean basin, was discovered in Australia around 1896 and has increased in importance with the establishment of improved pasture. Lucerne flea will attack cereal crops and both newly-sown and established legume pastures, including annual medics, sub clovers and lucerne.

Description

Lucerne flea is a plump, wingless insect, yellowish-green in colour with or without black or brown markings, and about 3 mm long. It is a chewing insect, and is equipped with a "spring" which allows it to jump about 30 cm at a time.

Life cycle and biology

Eggs are laid in clumps of 60–80 eggs on damp ground, usually under the shelter of plants. Each egg is covered with a small amount of excreta consisting mainly of soil previously swallowed by the adult for this purpose. Eggs develop only when wet. The coating of excreta around the eggs ensures the best possible flow of moisture from the soil. The time taken for eggs to hatch depends largely on temperatures. Under autumn conditions hatching occurs after approximately one week, but at temperatures of 10°C it takes about one month. The newly hatched nymph is a miniature of its parent, no more than 1 mm long and yellow in colour. It grows through a number of stages before reaching maturity. The first autumn generation takes about four weeks. A second generation may be completed before lower temperatures reduce development, resulting in larger populations during the winter. Upon increasing temperatures in the spring, lucerne flea begin to regenerate quicker and increase in numbers. When temperatures reach 15°C soil moisture conditions generally limit the development of eggs. Newly laid eggs die, while partly developed eggs that have undergone more than half of their development become dormant and are carried over until

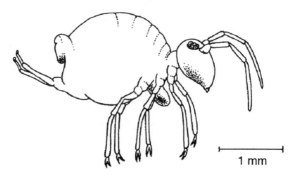

Figure 5.4 Lucerne flea (*Sminthurus viridis*)

the following autumn. A long wet spring allows a large build up of the dormant egg population, the effect being realised in the following autumn when large numbers of lucerne flea may hatch.

Plant damage and symptoms

Lucerne flea causes damage by chewing the green parts of leaves, leaving an exposed "membrane" or epidermis. Affected leaves have characteristic "windows" in them. Minute black specks of excreta are also present. Lucerne flea are often found in conjunction with red-legged earth mite, although the damage caused by each insect is quite different.

Control

Omethoate (Le-mat®), Technical Maldison (Malathion ULV®) and Chlorpyrifos (Lorsban®) are recommended for control of lucerne flea. Check current recommendations for rates and current chemical registration.

A predatory snout mite, *Biscirus lapidarius*, has been introduced as a biological control to help control lucerne flea. However, this mite does not reduce flea numbers sufficiently and insecticides are still required. Another mite predator (along with one for red-legged earth mite) was introduced into Western Australia from the Mediterranean basin during the 1960s. Although both have established, neither lucerne flea nor red-legged earth mite populations have shown any marked reduction.

The annual medic varieties of Paraponto, Sapo and Paragosa gama medic (*Medicago rugosa*) have some tolerance to lucerne flea.

Pasture cockchafer (black headed)
Aphodius tasmaniae

There are at least three other species of pasture cockchafer in southern Australia, namely *Adoryphorus couloni* (red headed) and *Sericesthis geminata* and *Sericesthis planiceps* (both Yellow headed). The following information applies to *Aphodius tasmaniae* (black headed).

Host range and distribution

The pasture cockchafer is a native insect which has been recognised as a pasture pest for the past 30–40 years. It is a pest of pastures in the higher rainfall areas, but occasionally attacks cereal crops.

Description

The adult is a dark brown to black beetle 10 mm long, has parallel sides and a broad shovel-like head. The abdomen, thorax and head are more or less the same width, and the head in particular is broad and shovel-like and is used to dig into the ground. The larvae have six legs with a dark brown to black head capsule, and grow to approximately 20 mm in length. Body colour is grey when they are feeding, but changing to a yellowish-white colour before pupation. The eggs are pale yellow spheres about 2 mm in diameter and are laid in batches at various depths in the soil. Larvae are grey in colour, with black or dark brown head capsules.

Life cycle and biology

There is only one generation of pasture cockchafers each year. During January-March the beetles fly in large numbers on warm nights and are attracted readily to lights. They are stimulated into activity by rain from mid-January onwards. Sufficient rain is required to penetrate about 150 mm into the ground, depending on soil type. Two to three days after the rain, beetles move on to the soil surface and if the temperatures are warm, the beetles fly, mate and lay their first batch of eggs. The beetles mate on the soil surface if temperatures are cooler, and then burrow back into the soil to lay their first batch of eggs, thereby having the ability to reinrest old cockchafer patches.

The beetles live for about two weeks. During this time they can lay two batches of eggs. The first batch of 30–50 eggs is laid about six days after the rain. They must then feed on dung for a few days before they can lay the second, smaller batch of about 10–25 eggs. Young grubs hatch after three to four weeks, and do not move far until it rains. After the rain, larvae come to the surface, making vertical tunnels in the soil and begin to feed on organic matter on the surface. Larvae must feed soon after hatching in order to survive. If rain does not fall for 30 days after hatching, 50 per cent of the larvae will die. However, young larvae that feed on organic matter soon after hatching are much more tolerant to dry periods.

There are three stages of larval development (called instars). First instar larvae feed on organic matter on the soil surface. Second instars have usually developed by May, and feed on green plant material. Most pasture damage is caused by the third instar larvae, which feed on plants that lie close to the ground or which have slender stems.

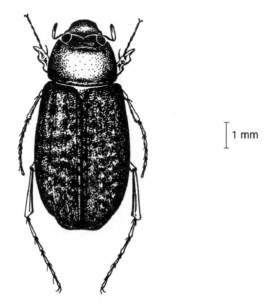

Figure 5.5 Pasture cockchafer (*Aphodius tasmaniae*)

From July onwards the larvae begin to store fat, and the body colour changes to a creamy white. In late August-September larvae seal themselves into cells at the bottom of their tunnels and in late December they pupate. Three weeks later new generation beetles will emerge when it rains.

Plant damage and symptoms

Third instar larvae chew through plant stems at ground level and usually take the cut portions into their burrows to feed. As food becomes scarce, larvae will move across the soil surface and construct new tunnels. Larvae can move up to four metres from their original tunnel in a season.

Control

Pasture cockchafers can be controlled by cultural techniques as well as with insecticides. As the beetles are attracted to bare areas, control can be achieved by maintaining a grass cover during the summer. This requires the incorporation of pasture improvement with grazing management. First instar larvae feed only on surface organic matter, and consequently can be controlled by removing the feed source by ploughing. This operation has to be carried out usually before the opening rains to be effective.

Lindane was an insecticide used to control pasture cockchafer, but it is no longer recommended. Check current recommendations for alternatives.

There are no available biological contols or resistant varieties.

Pea aphid
Acyrthosiphon pisum

Host range and distribution

The pea aphid is suited to cooler temperatures and occurs in association with blue-green aphid, although generally in much lower numbers. Pea aphid has a considerably broad host range, and is known to attack annual medics, sub clovers and other *Trifolium* spp. and many grain legume crops such as peas, beans, lentils and lupins.

Life cycle and biology

Pea aphid is found only as asexually reproducing females in Australia. Optimum multiplication occurs at temperatures of about 20°C, which means the pea aphid slots in between blue-green aphid and spotted alfalfa aphid, which have optimum reproducing temperatures of 18–20°C, and 20–25°C respectively.

It appears from field experience that pea aphid will be a problem in late winter and spring. Pea aphid looks much the same as blue-green aphid, although on some hosts it is slightly larger. Confident identification can only be undertaken using a low-powered microscope.

Plant damage and symptoms

Pea aphid damage is much the same as that of blue-green aphid, although plants will be completely killed by pea aphid. It is difficult to differentiate

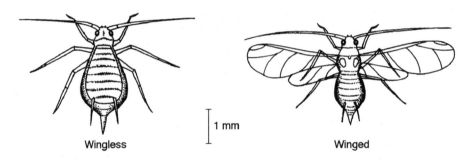

Figure 5.6 Pea aphid (*Acyrthosiphon pisum*)

between pea aphid and blue-green aphid in the field, and their associated damage. Recommended insecticides are the same for both insects.

Control

Pirimicarb (Pirimor®) and Demeton-S-methyl (Metasystox®) are recommended for use in annual medics, sub clovers and lucerne. Check current recommendations for other crops and rates, and current registration.

At this stage there is no parasite suitable for release to control pea aphid.

The only annual medic variety with tolerance to pea aphid at this stage is Harbinger strand medic (*Medicago littoralis*). Considerable research is being undertaken by the South Australian Research and Development Institute, and all lucerne varieties released by them have some degree of tolerance to pea aphid.

Pink cutworm
Agrotis munda

Host range and distribution

Pink cutworm is a native insect of Australia which causes serious damage in lucerne and summer fodder crops, and late sown annual medic and sub clover seed crops. It will also attack cereal and oilseed crops, young grapevines and some spring sown fodder crops throughout cool and warm temperate zones.

Description

The moths are about the same size and shape as climbing cutworm (*Helicoverpa punctigera*) moths, but the forewings are grey to brown in colour and distinctly but irregularly patterned. Young larvae are greyish green or brown in colour with a slight pink tinge and about 25 mm long. Older larvae are pink, 40 mm in length and are generally fat and sluggish.

Life cycle and biology

Moths first appear in late August and September after emerging from pupae that over-winter in the soil. Eggs are laid in the top 15 mm of soil, often

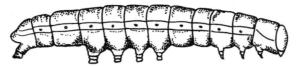

Figure 5.7 Pink cutworm (*Agrotis munda*)

near the base of plants. If the soil is very wet, eggs are laid on to the plant foliage. Females generally select relatively bare ground to lay eggs. The newly-hatched larvae first appear in mid-October to early-November and they are very active and disperse rapidly before feeding.

Larvae generally feed by night and shelter during the day at the base of the plants. Feeding generally ceases during December when larvae burrow into the soil, form an earthern cell and pupate. Occasionally a second or partial second generation occurs, but is generally restricted to the cooler high rainfall areas.

Plant damage and symptoms

Heaviest infestations generally occur in years where there is a dry spring and early summer. When newly hatched, young larvae climb plants and feed on the lower surfaces of the leaves. They generally leave the upper epidermis intact, producing a "window" effect by removing the green plant material but leaving the leaf veins. As they grow the larvae eat larger parts of the leaf and will completely strip the leaves.

Seedlings may be eaten off at ground level and large autumn sown plants may be ringbarked. In some instances established lucerne, particularly after the first hay cut, has been eaten as fast as it is produced.

Control

It is very important to assess larvae numbers, especially in the spring. Spring sown lucerne and summer fodder crops should be sprayed if one larvae per plant can be found.

Endosulfan (Thiodan®) and Chlorpyrifos (Lorsban®)are recommended but check current recommendations for alternatives and rates. Check current registration of chemicals.

There are no available biological controls or resistant pasture varieties.

Red-legged earth mite

Halotydeus destructor

Host range and distribution

Red-legged earth mite is an introduced pest that attacks crops and pastures throughout the warm and cool temperate zones.

Description

The fully grown mite is approximately 1.5 mm long, has a velvety black body and eight red legs. The mouthparts of red-legged earth mite are so structured that it will suck only sap. Red-legged earth mite occurs only as an asexually reproducing female. Young newly hatched mites are approximately 0.25 mm long.

Life cycle and biology

The red-legged earth mite lays eggs throughout the autumn and winter months, but towards the end of spring, mites cease laying eggs and begin to accumulate eggs inside their bodies. Upon death, a mite may contain up to 100 viable eggs. The dead bodies are scattered by wind and lodge in debris or small cracks in the soil surface. The mite bodies gradually disintegrate, leaving eggs which are highly resistant to drying out, and will survive hot summer temperatures with little or no protection. Over-summering eggs readily absorb moisture but require contact with moisture and temperatures below 16°C to begin development. This mechanism protects eggs against hatching in the summer when there could be a lack of food. Eggs take approximately two weeks to hatch, depending on temperatures. During this time they must remain in contact with moisture. Partial drying at this stage will either delay or permanently stop further development. Good opening rains and a follow-up will produce many hatchings. After hatching, mites are able to reproduce in about three weeks. Eggs laid during the autumn and winter months are very similar in appearance to the resistant over-summering eggs.

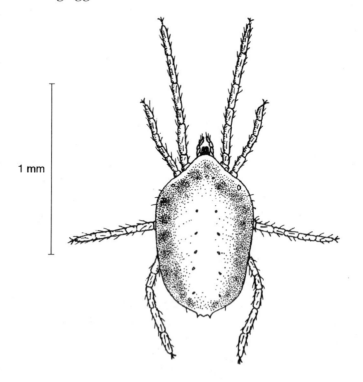

Figure 5.8 Red-legged earth mite (*Halotydeus destructor*)

Winter eggs are laid underneath plants and begin to develop immediately and may hatch in four days, demonstrating the potential to produce a second generation. The actual number of generations in a year varies depending on the environment. Moisture stress and lack of feed can reduce the number of generations and egg viability, thus having a significant effect on subsequent populations in the following year.

Plant damage and symptoms

Red-legged earth mite damage generally appears as a silvery discolouration on green plant tissue, caused by the removal of sap from the affected area. Damaged areas eventually turn brown, wither and die. The extent of plant damage depends upon mite populations and the growth and vigour of plants. Legume seedlings are very vulnerable to red-legged earth mite damage in the autumn.

Control

Omethoate (Le-mat®) and Chlorpyrifos (Lorsban®) are recommended for control of red-legged earth mite. Check current recommendations for rates and current registration of chemicals.

A mite predator was introduced into Western Australia from the Mediterranean basin during the 1960s. Although it has established since its introduction, red-legged earth mite populations have not reduced significantly.

The South Australian Research and Development Institute is undertaking breeding for resistance in annual medics and lucerne.

The annual medic varieties Sava snail medic (*Medicago scutellata*) and Sapo and Paraponto gama medic (*M. rugo*sa) have some degree of tolerance to red-legged earth mite.

Sitona weevil
Sitona discoideus

Host range and distribution

Sitona weevil was discovered in south-eastern Australia in the late 1960s and it was not long before its effect on pasture legumes, particularly annual medics in the warm temperate zones, was evident. Sitona weevil will also attack lucerne but all sub clovers are not affected in the field.

Description

The adult weevil is about 5 mm long and is dark greyish-brown with three white stripes on the thorax. The larvae of sitona grow up to 5 mm long, are white in colour, and occur in the top ten centimetres of soil.

Life cycle and biology

Sitona weevil undertakes one generation each year, where the adults invade newly regenerated pastures in autumn after moving out from sites under trees and in litter, where they spent the summer. After attacking the

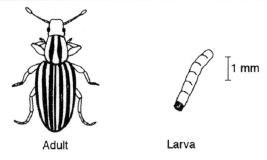

Figure 5.9 Sitona weevil (*Sitona discoideus*)

newly emerged annual medic plants, the adults mate and the females lay eggs in the undisturbed soil. A single female can lay up to 2000 eggs, which then hatch into small larvae. The larvae pupate in the early spring, and then emerge as sexually immature adults which attack the plant herbage. When plants dry off, the adults fly into litter under trees, for aestivation over the summer months.

Plant damage and symptoms

Adults attack the plant herbage, firstly by chewing around the outer perimeter of the leaves, leaving a scalloped appearance. In severe cases the complete leaf is eaten.

Larvae attack the nitrogen fixing nodules which are attached to the plant roots, and will also chew through roots, thus inactivating a large root and nodule network.

Control

It is extremely hard to control sitona weevil with insecticides due to their hard shell and also because continuous flying-in of adults from other areas occurs. In severe cases, the use of Fenitrothion (Folithion 50®) or Azinphos-Ethyl (Gusathion A®) is recommended. Check current recommendations for rates. It is not possible to control sitona larvae with insecticides at this stage.

The South Australian Research and Development Institute has released a sitona adult wasp, *Microctonus aethiopoides*, which was introduced from Morocco. It is approximately 3 mm long, the females being brown and the males black. It has been found quite some distance from its release sites and has been recovered from sitona adults, which proves its potential in an integrated control approach for sitona weevil control.

Considerable research has been undertaken by the South Australia Department of Agriculture in breeding for resistance to sitona weevil adult. Sava snail medic (*Medicago scutellata*) and Sapo gama medic (*M. rugosa*) have tolerance to sitona adult. (That is, sitona adults will still attack these varieties, but to a lesser extent than a completely susceptible plant.)

Current research is being undertaken to breed resistance to sitona weevil larvae.

Spotted alfalfa aphid
Therioaphis trifolii

Host range and distribution

Spotted alfalfa aphid was discovered in eastern Australia in March 1977. It is a pest of legumes in the cool and warm temperate zones, is more suited to higher temperatures, and has caused major destruction in lucerne stands (both irrigated and dryland). It can cause damage to annual medics in the early autumn and late spring when temperatures are around 20–25°C. All sub clovers are either very tolerant or resistant to spotted alfalfa aphid.

Description

Spotted alfalfa aphid is a small yellow-green aphid with four to six conspicuous rows of dark spots with spines on the upper abdomen. Adults are approximately 2 mm long. The winged form has smoky marks around the veins in the wings.

Life cycle and biology

The aphid normally occurs as an asexually reproducing female, both in the winged and wingless form. More winged aphids develop upon overcrowding. Optimum reproductive temperatures are 20–25°C and 25–30 per cent relative humidity. 20–40 generations can occur each year.

Plant damage and symptoms

Early spotted alfalfa aphid attack is recognised as vein clearing on the leaves. An easy check is to shake a few plants on to the hand or a piece of paper.

Spotted alfalfa aphid's mechanism of attack is by the sucking of sap from the plant, together with the injection of salivary toxins into the plant. Secondary damage occurs when large quantities of honey dew (excreted excess sugars) cause prolific growth of moulds. Spotted alfalfa aphid will stay on the plant until it is completely dead.

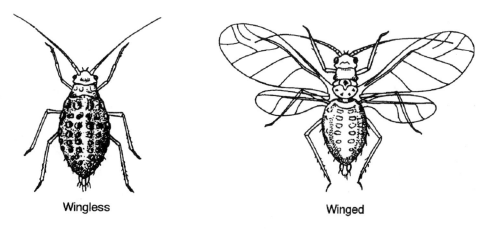

Figure 5.10 Spotted alfalfa aphid (*Therioaphis trifolii*)

Control

Insecticide treatment is as for blue-green aphid. Check current recommendations for rates, as too high rates may kill important predators and spotted alfalfa aphid parasites, and registration.

The spotted alfalfa aphid parasite, *Trioxys complanatus* is a small wasp which was introduced and released in 1977. The parasite was mass-reared and released on to the properties of collaborating farmers, particularly in lucerne stands. It has proven to be an important spotted alfalfa aphid control agent. The common ladybird is also a predator of spotted alfalfa aphid.

Considerable research has been undertaken to develop spotted alfalfa aphid resistance in annual medics and lucerne. All varieties resistant to blue-green aphid also have resistance to spotted alfalfa aphid.

6

Weeds in pastures

Before considering any aspect of weed control in any situation it is necessary to have a clear understanding of why some plants are called weeds. Traditionally a weed is defined as "a plant growing out of place" or "an undesirable plant". Many plants classed as weeds were once valued plants of home gardens, soursob, horehound, and cape tulip are good examples of garden escapees (Figure 6.1). Valued pasture plants growing where they are not wanted can also be classed as weeds. Common examples are annual ryegrass in cereal crops and phalaris on roadsides (Figure 6.2).

In cropping situations it is relatively easy to define a weed as practically any plant other than the sown species. In pastures, however, the situation is not as simple. Pastures usually contain a variety of species and it is sometimes a matter of opinion as to whether some species are weeds or merely relatively poor pasture species of little value.

Basic considerations when defining weeds in pastures are:
1. Competition. All non-pasture plants growing within a pasture are competing with the pasture species for moisture, nutrients and sunlight, weakening the growth of the pasture, reducing its potential yield and lowering its feed value. These non-pasture species should be regarded as weeds. Common examples are capeweed, storksbill and barley grass (Figure 6.3).
2. Toxicity. Any plant that is toxic to animals must be classed as undesirable. Examples of weeds that are toxic to livestock are cape tulip, salvation Jane (Patterson's curse) and potato weed (Figure 6.4).
3. Palatability. If livestock will not eat particular plants then they serve no useful purpose in a pasture and are wasting space and causing stock to graze out the more palatable species.
4. Productivity. Plants that are relatively unproductive because they either have a short growing period or just lack vigour lower the overall production of pastures and may be considered weeds.

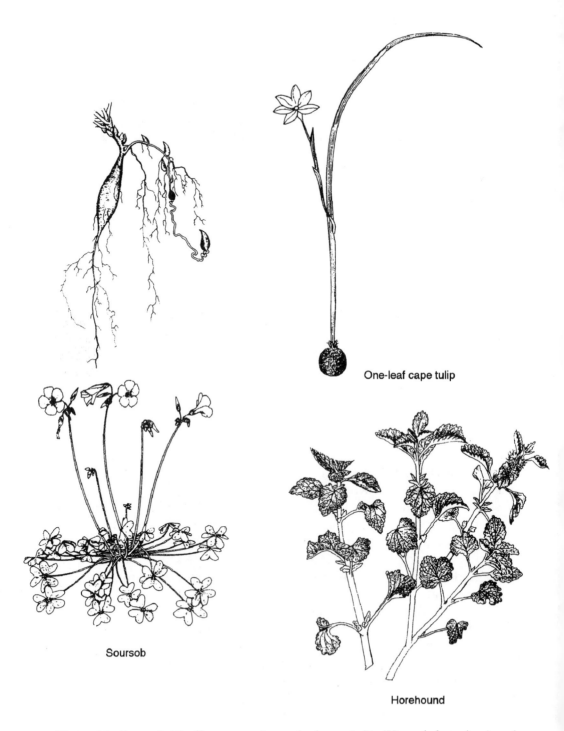

Figure 6.1 Soursob (*Oxalis pescaprae*), one-leaf cape tulip (*Homeria breyniana*) and horehound (*Marrubium vulgare*)

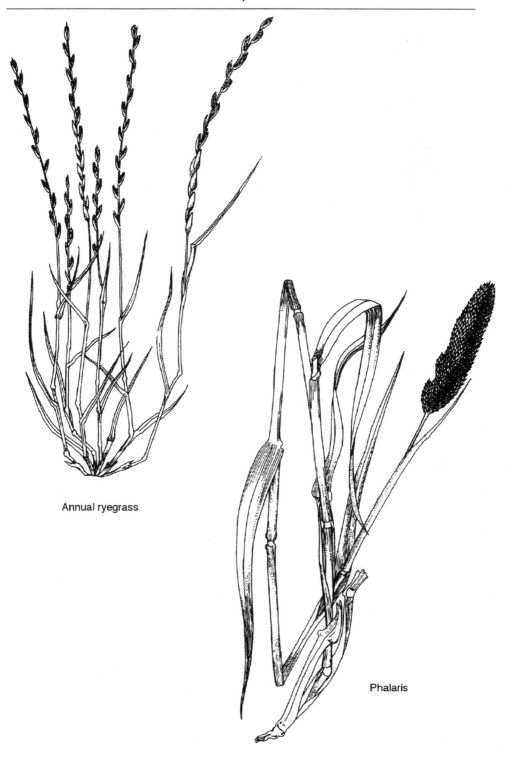

Figure 6.2 Annual ryegrass (*Lolium rigidum*) and phalaris (*Phalaris aquatica*)

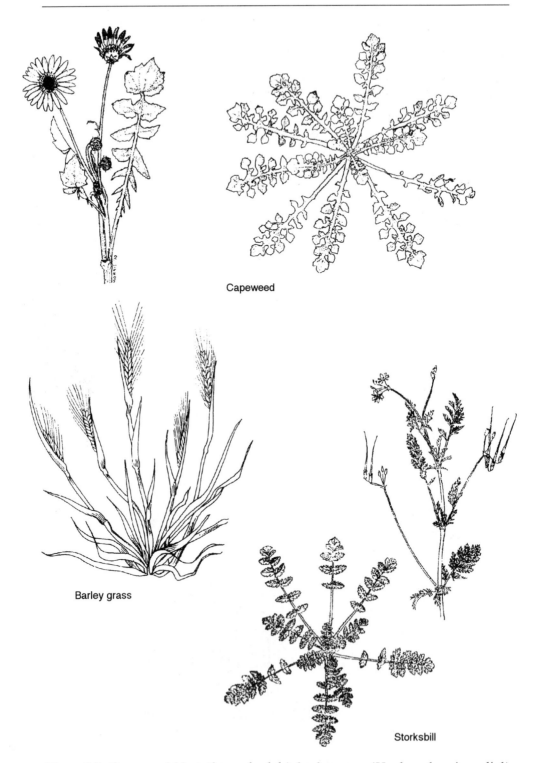

Figure 6.3 Capeweed (*Arctotheca calendula*), barley grass (*Hordeum leporinum link*) and storksbill (*Erodium* spp.)

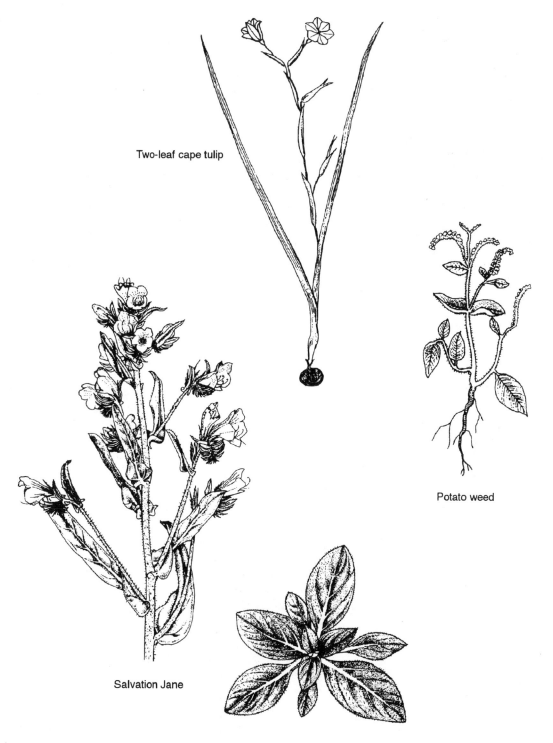

Figure 6.4 Salvation Jane (*Echium plantageneum*), potato weed (*Heliotropium europaeum*) and two-leaf cape tulip (*Homeria miniata*)

Pasture management

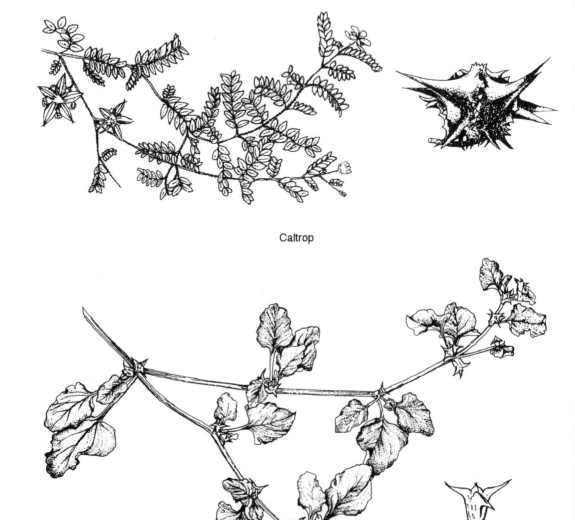

Figure 6.5 Caltrop (*Tribulus terrestris*) and three cornered jack (*Emex australis*)

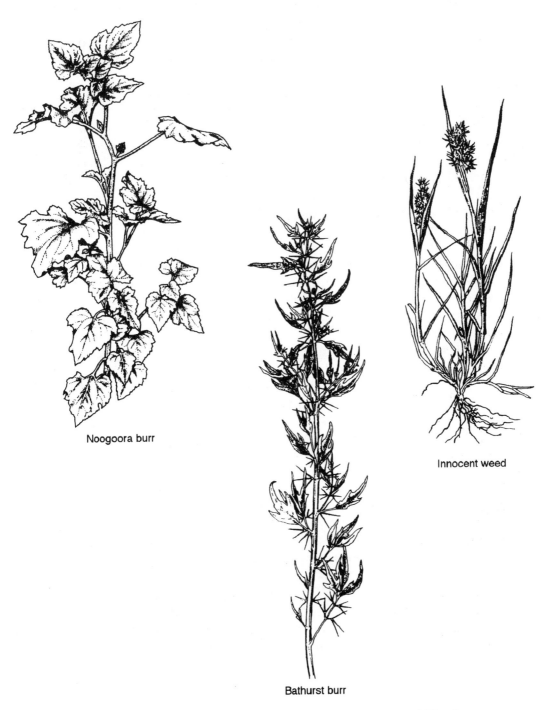

Figure 6.6 Noogoora burr (*Xanthium pungens*), innocent weed (*Cenchrus pauciflorus*) and Bathurst burr (*Xanthium spinosum*)

5. Mechanical injury. Some plants cause mechanical or physical injuries to livestock or damage livestock products such as wool and hides. Examples of such weeds are caltrop, three-cornered jack, Noogoora and Bathurst burrs and innocent weed (Figures 6.5 and 6.6).
6. Host disease organisms and harbour pests. Plants that act as hosts for diseases of wanted plants such as cereals, or harbour vermin must be considered weeds in pastures.

All these factors detract from productivity and must be taken into account when managing pastures and assessing the relative importance of particular weeds.

It is also necessary for all landowners to be fully aware of their legal responsibilities in respect to the control of noxious weeds that have been proclaimed pest plants under various state legislation.

The legislation is designed to protect the community from financial loss or nuisance due to serious weeds, and aims to eradicate some weeds and control others.

Prevention is better than cure is an old adage that applies to the introduction of weeds as well as illness.

Weeds do not just appear; they are introduced onto properties and into pastures in many different ways. Any traffic entering a property must be regarded as a potential source of contamination, and steps taken to minimise the risks.

Vehicles, machinery, farm produce and livestock entering a property should all be inspected for cleanliness and if necessary held and cleaned in a quarantine area prior to use if suspected of being contaminated.

Many weed seeds readily stick to wool and the coats of animals or clothing, some withstand the digestive processes of livestock and are transported in the animals gut and deposited in dung, while others lodge and stick in any small crack or crevice in machinery. Constant care is needed to prevent weed seeds being spread in this way.

Care should also be taken when buying pasture seed and fodder, such as hay, to ensure that weed seeds are not accidentally introduced in this way. Purchase certified seed, and inspect the certificate of analysis to ensure that pasture seed is weed free, and always purchase hay from known weed free sources. If in doubt inspect the paddocks from which the hay is to be cut prior to purchasing.

All unusual plants found in pastures should be regarded with suspicion as potentially serious weeds and every effort should be made to identify them quickly so that steps can be taken to control or eradicate the weed before it has a chance to become established.

Weed control methods and techniques

Weeds can be controlled in many different ways. The method or technique chosen to control a particular weed will depend on its growth habits, stage of growth and the situation in which it is growing. The method by which the weed reproduces also influences the selection of effective control measures.

To enable effective control it is necessary to categorise weeds into two main groups according to their life cycles.
1. Annuals. These are plants that complete their life cycle within one year. They rely entirely on seed production for their regeneration the following year. Annuals are prolific seeders and since most weed seeds are able to remain viable in the soil for many years, there may be huge reserves of seed that need to be exhausted before any significant reduction can be seen in the weed population. It is therefore extremely important to prevent annuals from setting seed as the seed set in any one year will ensure that reinfestation occurs for many years to follow.
2. Perennials. These are plants that live for more than two years. Some reproduce only from seed, but many also have the ability to grow from vegetative parts such as rhizomes, stolons, bulbs and tubers as well as from seed. To control these effectively it is therefore necessary to kill all the vegetative parts, both above and below ground.

Control methods will therefore differ for annuals and perennials and for different perennials. Control methods that are suitable for annuals may actually help to spread some perennial weeds.

It is essential that you have an appreciation of these different groups in order to select an appropriate and effective means of control.

Methods of control can be put into the following categories:
1. Cultural
2. Biological
3. Chemical.

They are all useful but usually a combination of different methods is required for effective control, especially long term control.

Cultural control methods

Pasture weed control must be planned and started at least one year before establishment or renovation. This is particularly so in "old" country previously sown to crop or pasture. Control measures must start by preventing existing weeds from setting seed, and by removing excessive vegetative growth by either grazing or slashing.

Cultivation is the oldest method of control and carried out at the correct time will eliminate many weeds, particularly annual weeds, before the pasture is sown. Cultivations should be timed to kill as many weeds as possible following germination, and repeated as necessary to obtain a clean weed free seed bed.

Shallow cultivations are more useful for weed control than deep ploughing. Deep ploughing tends to bring buried weed seeds to the surface where they can germinate and also bury other weed seeds. The infestation of new pastures by weeds that have not been seen on a property for years is sometimes due to deep cultivation.

While cultivation is an excellent method of controlling annual weeds it may actually help to spread many perennial weeds. Cultivation should be avoided as much as possible where there are perennial weeds that grow from rhizomes, stolons, bulbs, corms and tubers as it only serves to distribute them and increase the size of the infestation. Chemical control is usually required in these situations.

Mowing, slashing and grazing are useful methods for preventing annual weeds from seeding. They can also be used to reduce the vigour of perennials by depleting their food resources, but they must be followed up by chemical control.

When considering grazing make sure that the weeds are palatable and non-toxic, otherwise stock will avoid them in preference to more palatable pasture species and the problem will get worse. The pasture species will be over grazed and the weeds encouraged by lack of competition.

Competition from desirable plants is essential if weeds are going to be controlled in pastures. Weeds cannot grow in a place that is already occupied by a healthy pasture plant. Consequently high seeding rates combined with adequate applications of fertiliser, and good grazing management, to prevent both over grazing and under grazing, and disease and insect control will do much to establish and maintain a dense pasture sward that will crowd out weeds and prevent them from establishing.

Burning is a method of control that is seldom used in pastures except to remove excess trash to enable more effective cultivation, or to break the dormancy of bulb weeds such as Cape tulip to encourage a high germination percentage which in turn will enable a more effective chemical control than would otherwise be possible.

Cleaning crops are an option where there are serious weed infestation problems and it is apparent that you are not going to be able to control the weeds in one season either by cultural or chemical control methods. It is often more economical to sow a cereal crop such as oats, wheat or barley the year before sowing a new pasture.

Cereal crops are particularly useful cleaning crops as many problem broad leaf weeds can be controlled with herbicides in the crops without damaging them. Thistles, docks (Figure 6.7), crucifers, salvation Jane, capeweed and storksbill are some of the weeds that can be controlled by applications of selective herbicides such as 2, 4-D, MCPA or dicamba. One year's cropping incorporating effective weed control measures followed by a short autumn cultivation before sowing the pasture should satisfactorily control most common pasture weeds. In some cases where dock infestations have been serious, two successive cereal crops may be needed to control the weeds.

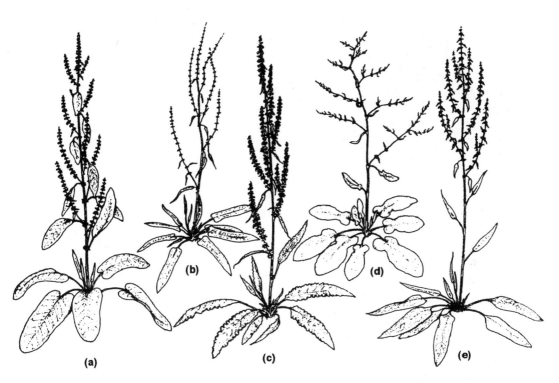

Figure 6.7 Dock species. **(a)** Broad-leaved dock (*Rumex obtusifolius*), **(b)** Swamp dock (*Rumex brownii*), **(c)** Curled dock (*Rumex crispus*), **(d)** Fiddle dock (*Rumex pulcher*) and **(e)** Clustered dock (*Rumex conglomeratus*)

Where irrigation is available, spring sown crops such as millet, hybrid maize, and the sorghum/Sudan grass hybrids, such as Speedfeed and Sudax, can be used as cleaning crops in the same manner as autumn sown cereals. However, the spring sown crops are not as effective as autumn sown crops for controlling dock.

Biological weed control

This method involves using other living organisms to control weeds. The biological agents are usually small forms of life such as insects or fungi.

This type of control depends on finding suitable biological agents that will not only persist and be self regenerating in the environment in which the weed is growing and also be safe to use, not causing damage to other useful and wanted species. Finding such agents involves extensive research, often in the country from which the weed originated, a great deal of expense, and no guarantees of success.

In Australia this type of research is carried out by organisations such as the CSIRO and various universities.

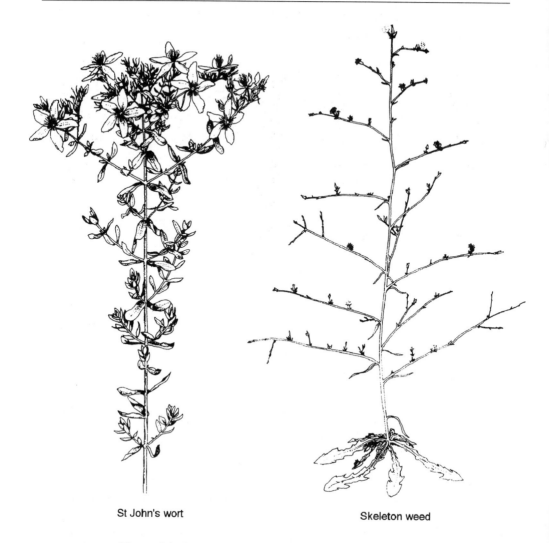

Figure 6.8 St John's wort (*Hypercium perforatum*) and skeleton weed (*Chondrilla juncia*)

Biological control organisms will not completely eradicate a weed. They will, however, reduce the size, density and vigour of infestations to a level where other more conventional methods of control can then be used successfully and economically.

Biological agents have been released in an attempt to control a range of weeds, some of which are prickly pear, Noogoora burr, St John's wort and skeleton weed (Figure 6.8). The success of these cases has varied from spectacular, in the case of prickly pear, to failure in the case of St John's wort.

Research into biological control agents is a continuous and time consuming process. It is possible that in the future new biological weed control agents will be released and be successful, however it would be unwise for a farmer to ignore weed problems in the hope that biological control strategies will be implemented.

Chemical control

Note that the information below should only be considered a broad guide. Some of the chemicals mentioned may have been deregistered in your state or may become so. You should contact local authorities for information and current recommendations.

Remember also to READ THE LABEL of any chemical you are using or about to use. It is illegal to use a chemical for a purpose or in a manner other than those stated on the chemical's label.

Whenever weed control is mentioned many people immediately think that chemicals or herbicides are the answer to the problem. Unfortunately the answer is not so simple.

Herbicides are only one tool to use in an overall weed control program. Unless they are used in conjunction with the other control methods already mentioned they too will represent only a partial answer and a temporary measure.

Chemical weed control can be divided into two main types:
1. Total control
2. Selective control.

Total weed control involves killing all the plant life in a certain area and, depending on the herbicides used, this may be for only a short time or for a period of up to several years.

Situations where long term total control is required include along fencelines, firebreaks, roadways, storage areas and on irrigation banks and ditches. Short term total control is required in pastures for small patches of noxious weeds and for a complete kill of weeds prior to sowing or sod seeding.

Herbicides used for total control usually consist of a combination of two or more chemicals designed firstly to give a quick kill or knockdown of existing growth and secondly to give residual control and prevent subsequent germinations. There are a number of such herbicides available commercially. The product selected and the rate used will vary with the situation and the required results. The most common herbicides used are amitrole, atrazine, diuron, glyphosate and simazine in different combinations. Amitrole and glyphosate are used to kill existing growth and atrazine, diuron and simazine to provide residual control.

Small patches of noxious weeds discovered in pastures should be treated as quickly as possible with a total weed killer. The infestation should be isolated and clearly marked to prevent livestock distributing seed and vegetative parts, and to enable easy and regular observation. Any flowering parts should be carefully removed, bagged and then burnt. Glyphosate should be used to control this sort of infestation where residual control is not required or a total soil sterilant such as picloram can be used. It may be necessary to keep such areas under observation for several years to ensure further infestations do not occur.

A complete kill of weeds prior to sowing can be obtained using desiccant type herbicides which when applied to the foliage give a quick knockdown in just a few days. This is a practice that is becoming increasingly popular for a number of reasons. Besides being economical it avoids long fallows so paddocks are not out of use for more than a few weeks prior to seeding, and it also reduces the risk of soil erosion. It is also quick and particularly useful in wet seasons when it is difficult to use normal cultivation equipment.

The most commonly used herbicides for this type of operation are diquat (Reglone®) for broad leaf weeds and paraquat (Gramoxone®) for grass weeds (such as silver grass, Figure 6.9), both of which are desiccants. Sprayseed® is a mixture of both diquat and paraquat to achieve a broad spectrum control.

Figure 6.9 Silver grass (*Vulpia* spp.)

The most important points to remember when using these herbicides are:
- Use the recommended quantity of water per hectare as good coverage is essential
- Follow all directions on the herbicide label
- Mix with clean water as these chemicals are deactivated by clay particles in dirty water
- Only spray seedling growth. Higher application rates must be used on more mature growth and results may vary.

Glyphosate (Roundup®) which is a foliar-applied translocated herbicide can be used to do the same job. It will kill perennial weeds as well as annuals, but because of its translocated mode of action it takes longer to act. Weeds sprayed with glyphosate may take several weeks to die.

Selective weed control

This in essence means killing the weeds while leaving the desirable pasture species unharmed. In pastures the phenoxy acetic acid group of herbicides are the most commonly used. Most other herbicides that could be used are uneconomic in this situation. This group of herbicides are only active on broad leaf plants and since most of our weeds fall into this category this limitation is not very important. Clovers and medics are of course broad leaf plants and as a result are prone to damage from the more severe forms, and from excessively high rates of the less severe forms. The most common forms of phenoxy acetic acid herbicides listed in order of severity to clovers and medics are as follows:

2,4-D Ester	Not recommended on clovers or medics.
2,4-D Amine	Safe at very low rates, medics are more susceptible than clovers.
MCPA	Safe on both clovers and medics.
2,4-DB	Safer than MCPA but because of cost usually only used in the establishment year. Safe in seedling lucerne.
MCPB	The safest on both clover and medics, but the most expensive and consequently seldom used except in special circumstances.

For general broad leaf weed control in pastures, Amine 2,4-D and MCPA are our most useful herbicides and at rates of 700–1400 mls/ha will control most weeds. The higher rates must be used on older weeds.

In mature lucerne pastures Amine 2,4-D, Ester 2,4-D and MCPA may be used provided the pasture is first grazed down to ground level. Lucerne is susceptible to these herbicides, consequently to reduce the likelihood of damage, spraying should be confined to the winter dormant period (preferably no later than the end of July) and heavily grazed to remove surface growth prior to spraying. In seedling lucerne 2,4-DB can be used with safety from the first to the eighth trifoliate leaf stage. After this stage damage can occur.

Diquat and paraquat are useful for removing seedling weeds early in the season to allow pasture species to get established. Pasture will tend to smother later germinations. Diquat is used for the broad leaf weeds such as capeweed, and paraquat for grassy weeds such as barley grass. Perennial pasture species will be damaged by this treatment but will recover fairly quickly.

It is important to check for the most current registrations of herbicides and their recommendations.

Timing of herbicide treatment is of prime importance. The best time to spray is while weeds are still in the seedling stage and before the flowering stem emerges, for the following reasons:

- The weed has not yet begun to cause damage or appreciably compete with pasture species
- The weed is at its most susceptible stage
- Less herbicide is required on young plants
- Since lower rates of herbicide can be used, pasture species are less likely to be damaged.

Wetting agents should be used as stated on the label of the herbicide.

Spray-graze technique

This technique of controlling weeds is perhaps the most useful method of weed control in pastures, and can be used on a wide range of weeds provided the weeds are not poisonous. Many weeds such as capeweed and salvation Jane become dominant in pasture because they are less palatable than the desirable pasture species. The spray-graze technique relies on the fact that 2,4-D tends to raise the sugar content of plants and thus makes them more palatable. In addition, the leaves become more crisp and stand up, i.e. plants with leaves that lie flat on the ground such as salvation Jane are difficult to feed, 2,4-D makes these leaves grow upright and are thus more easily eaten. Low rates of herbicide are used sufficient only to give a partial kill of the weed and to produce these changes in the plant. Stock are used to complete the job.

An example of this technique with salvation Jane involves the following steps:

1. Remove stock.
2. Boom spray paddock with 50 per cent Amine 2, 4-D at 350–700 mls/ha. Use the low rate if plants are 5 cm or less across.
3. Withhold stock until the leaves curl up and become crisp. This takes between 7 and 14 days depending on the season.
4. Bring in sufficient stock to heavily graze the area to ground level in two weeks. Wethers are most suitable for this, but other sheep and cattle may be used equally successfully. Horses are not suitable.
5. Remove stock to allow pasture to recover.

This technique may have to be repeated for second germinations, however the low cost of the chemical involved makes this an economic proposition. The treatment must be repeated in following seasons so as to exhaust the seed reserves held in the ground.

Advantages of the technique include low cost of treatment plus the feed value achieved by grazing.

This technique can be used in lucerne provided the lucerne is grazed down to ground level first.

The success of the technique depends on achieving quick and even grazing. To do this paddocks should be no bigger than 20 hectares and grazing should be at the rate of approximately 50 sheep per hectare.

Ropewick application of herbicides

Ropewick application is a technique used to enable selective weed control using a non-selective herbicide, glyphosate.

The technique can be used only when the weeds to be controlled are at least 20 cm taller than the pasture they are infesting. This may necessitate grazing the pasture down low prior to application.

A ropewick applicator is basically a plastic pipe sealed at each end to act as a tank, with a nylon rope threaded in and out of it through a series of holes (Figure 6.10).

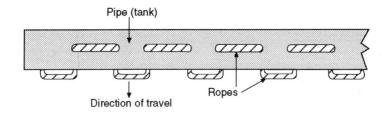

Figure 6.10 Pipe applicator

The pipe is filled with a concentrated solution of herbicide which keeps the rope constantly moist. The applicator is mounted horizontally across the front of a tractor so that when the tractor is driven forward the applicator brushes against the taller weeds wiping them with herbicide and leaving the pasture plants untouched (Figure 6.11).

Figure 6.11 Action of the ropewick applicator

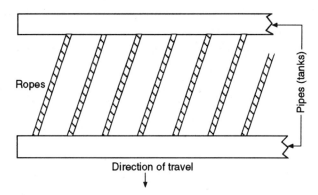

Figure 6.12 Multi-rope applicator

Multi-rope applicators can also be used in a similar manner, and because of their greater surface area they can be used at higher speeds than the pipe applicators.

Best results are obtained from ropewick applicators if the following rules are followed:
1. Graze pasture down hard.
2. Ensure weeds are growing vigorously.
3. Do not travel too quickly. 5–7 km/h (quick walking pace) is suitable for most situations.
4. Slow down for thick patches of weeds.
5. Treat all weeds twice, with the second application being applied in the opposite direction to the first.
6. Use a unit with a flow control valve to ensure the rope remains constantly moist even when the tank is nearly empty.
7. Keep ropes clean.
8. Use clean water to dilute the herbicide.

A hand held model of the ropewick applicators are weed wands which are ideal for spotting isolated weeds in pastures. This can be done without causing any damage at all to the surrounding plants.

Herbicides

Herbicides can be classified in a number of ways, but probably the most practical for our purposes is classification by their mode of action. In this way, herbicides fall into three main groups.

Contact

Contact herbicides are applied to the foliage and will only kill that part of the plant that the chemical comes in contact with. There is virtually no movement of the chemical within the plant. To achieve good results it is therefore necessary to have good coverage.

Herbicides that fall into this category include paraquat, diquat and bromoxynil.

Only seedling growth should be treated, older plants require higher rates of chemical in higher rates of water. The herbicides should be applied with a boom spray. Misting machines should never be used.

There is no residual life in the soil and the herbicides are non-volatile.

Translocated

Translocated herbicides are applied to the foliage and are able to move within the sap stream of the plant. Thus a foliage application will also kill the roots. Herbicides that fall into this category include the following:

- MCPA and 2,4-D. Known as broad leaf killers they can also affect grasses. Seedling grass and seed crops and pastures should not be sprayed for at least six weeks after emergence. Basically foliage translocated, these herbicides have some effect through the soil. Plant roots can pick up the herbicide, although at normal spray rates damage to established plants, particularly perennial, is negligible. Seed should not be sown for at least three weeks after spraying.

    Most MCPA formulations are sodium salts and as such are water soluble and most susceptible to rain after spraying. Do not spray when rain is likely within 24 hours; light rain after 12 hours probably has little effect.

    Ester and amine 2,4-D are less susceptible to rain after spraying. Ester (an emulsifiable concentrate) may be safe after six hours from spraying, but amine could require ten hours.

    In the choice between ester, amine and MCPA there are some guides to follow.

    As a broad statement MCPA does its best work against succulent annuals. Amine 2,4-D is best on similar or more fibrous plants. Ester 2,4-D is usually kept for more woody plants particularly perennials.

    The severity of the action of these herbicides against weeds also applies to susceptable valuable pasture species. This is amplified by the fact that esters are considerably more volatile than amine 2,4-D or MCPA thus increasing the risk of drift to valuable susceptable plants. It is therefore sound practice to use the more expensive low volatility formulations of esters near wanted susceptable species.

    Esters enter the plant more easily and rapidly than amine 2,4-D or MCPA but once inside the plant amine 2,4-D and MCPA translocate better than ester.

- 2,4-DB is a two stage plant killer. It must first be broken down within the plant to form 2,4-D before it becomes active, i.e. the plant causes its own death.

Some plants lack the ability to carry out this breakdown. Lucerne, clovers and medics lack this breakdown ability in the young stages. When clovers and medics are seedling plants of 2–5 true leaves and lucerne is a seedling plant of 2–8 true leaves spraying with 2,4-DB is safe. In both instances do not count the cotyledon leaves. However, before or after these plant stages the chemical is broken down and the application can damage similarly to 2,4-D.

- 2,2-DPA is an efficient grass killer (from couch grass to reeds and rushes), it is foliage translocated and has a residual life in the soil of about six weeks. At couch grass killing rates it will damage lucerne and many broadleaf pasture plants. It is essential to add a wetting agent. It is non-volatile. (This chemical has been largely superseded by glyphosate — a more efficient product.)

- Amitrole is an almost total killer in that it will kill or damage most annual broadleaf plants and grasses.

  This herbicide's killing action is to prevent the plant manufacturing food and for this reason it is not usually regarded as efficient on perennials. Perennials usually have enough food reserves to outlive the chemical. Low rates are frequently used on irrigation channels to suppress couch to a point where it does not interfere with water flow but retains enough foliage to prevent erosion.

  The herbicide has a residual life in the soil of about six weeks but its life within the plant may be considerably longer. Soursob and cape tulip bulbs can store the chemical and produce chlorotic plants the following year. It is non-volatile.

- Glyphosate is probably the most useful herbicide available for controlling a wide range of broadleaf and grass weeds, both annuals and perennials. Marketed in products such as Roundup® and Comkill® glyphosate is non-selective, non-residual, and non-volatile. It will only damage or kill the plants to whose foliage it is applied, and this makes it ideal to use for spot spraying patches of unwanted weeds in pastures, or in situations where volatile and residual chemicals cannot be used.

  Selective control can be obtained by applying glyphosate with a ropewick applicator which enables good control of tall weeds of pastures such as rushes, cape tulip, bracken, and salvation Jane.

  Glyphosate is deactivated by clay particles so care should be taken to use clean water when mixing this chemical.

Soil residual

Soil residual herbicides are applied to the ground either while the ground is bare, while the plants are still small seedlings or later in the plants growth often in combination with another chemical. Soil residual herbicides are basically absorbed into the plant through the roots although in many cases a limited amount of leaf absorption does take place. The efficiency and residual properties of these chemicals is largely dependent on their solubility and to their ability to adhere to soil particles. Very insoluble chemicals tend to remain in the top few centimetres or so of the soil while the very soluble chemicals will leach down very quickly. Chemicals in the triazine, e.g. simazine, atrazine, and substituted urea groups, e.g. diuron, linuron and chlorsulfuron, adhere tightly to soil particles, consequently a higher rate is needed on heavy soil than is necessary on sandy soils.

- Simazine is a soil residual herbicide with practically no leaf absorption qualities. Consequently it can be sprayed over the top of established growth without damage to the foliage. However, damage can occur through root absorption if the correct rates are not used. It is particularly useful on shallow rooted annual seedling growth, but it is not commonly used in pastures. It is non-volatile.
- Picloram plus amine 2,4-D known commercially as Tordon 50D®. It is an extremely potent chemical against lucerne, clovers and medics, most broadleaf plants as well as shrubs and trees. Tordon 50D® has an extremely long residual life in the soil. Three litres of Tordon per hectare will prohibit the growth of clovers and medics for over 12 months but this rate will not affect grasses. At the higher rates as would be used to kill skeleton weed it will kill clovers for approximately two years. As the chemical leaches down through the soil it will kill most shrubs and trees. Lucerne may not be grown for up to five years after spraying.

    Picloram is non-volative but the amine 2,4-D content is slightly volatile. From a drift angle it should be treated similarly to amine 2,4-D. The herbicide is useful in pasture situations to kill noxious weeds such as silverleaf nightshade, skeleton weed or hard to kill weeds that are still only a patch problem. It should never be used on broadacre basis except in very extreme cases.
- Bromacil is almost a total killer of broadleaf and annual plants. Its action is slow and, in fact, when used on boxthorn the plant may take 18 months or more to die. During this period leaves may form then wither and drop off on several occasions before the plant finally dies. The herbicide has a long residual life in the soil and at normal rates little growth can be expected for two years. It leaches very badly both down and sideways and should never be used anywhere near trees or valuable plants.

    Its use in pastures is confined to soil sterilisation of small patches of potentially troublesome weeds.

Wetting agents

Wetting agents or surfactants as they are more correctly called, are used to help spray solutions cover the plant and to stick the herbicide to the leaves.

In this way they are similar to a detergent, however, they are not detergents and detergents should not be used in their place for two main reasons. Detergents are ionic; that is they contain free electrically charged molecules, consequently, when added to an ionic chemical such as paraquat the chemical becomes inactivated (clay particles in dirty water will do the same thing). This is why a non-ionic wetting agent is always specified for most herbicide use e.g. paraquat and diquat. All reputable brands of wetting agent are non-ionic.

The second reason for not using a detergent is that detergents are "multiple wetting" while wetting agents are "single wetting". This means that detergent when first used will break down the surface tension of water and if that water dries up and the same area is wet a second time the detergent is still active and will once again break down the surface tension of the water. Wetting agents will only act on the water in this manner once.

Applied to herbicide application, if a detergent is used and then a light shower of rain falls within a couple of hours, the herbicide will be more easily washed off the plant. This does not happen with a wetting agent.

Wetting agents should not be used in selective situations unless specifically recommended, since the chemical will then stick to the desirable plants and damage can result. However, it may be used in all total weed control situations and for spot spraying, where more efficient absorption of chemical is required.

Application of herbicides

The effectiveness of a herbicide is largely dependent on how well and how accurately it is applied. If the weed does not receive the right amount of chemical or if the weed is not adequately covered by the spray then a poor kill will result. Conversely if too much chemical is applied unnecessary damage to pasture could result and of course herbicide is wasted and this means unnecessary expense.

The most common reason for herbicide failure is poor application.

Herbicides are applied by five main methods:
1. Boom sprays. Machines are the most accurate form of herbicide application available. A constant rate of spray solution at a constant pressure applied at a constant rate of travel through evenly spaced nozzles less than 30 cm from the ground results in an even distribution of herbicide over the area. All types of herbicides may be used with this type of equipment. The only disadvantage is that they are not suitable on hilly ground or where a very high rate of water is necessary. Normal application rates are in the range of 50 to 250 litres per hectare. If higher rates of water are needed it will be necessary to travel at a slower speed or increase the nozzle size and the pump capacity.

2. Cluster Jets. Where the use of a boom spray is not practical, e.g. in rough country, a cluster of jets (usually three) gives the approximate effect of a boom. They perform poorly under gusty conditions and do not force the spray into the foliage as effectively as a boom spray. They are not as accurate as a boom spray and once again variable results can be expected.
3. Hand spraying. This may be by means of an engine operated pump unit with one or two spray lines or a knapsack unit. Both types of equipment apply in the vicinity of 2500 litres of water per hectare; depending on the operator and the type of growth being treated. Dense growth and tall growth will use up a greater quantity of spray solution than will sparse and low growth.

   Hand spraying is the ideal method of application for spot spraying isolated plants and small patches where the infestation is not extensive enough to warrant the use of broadacre treatment. More severe types of herbicides may be used with this method since any damage that may take place on the pasture will be limited to those small areas sprayed.

   Recent advances in spot spraying technology have put light sensors on boomsprays to allow individual nozzles to operate only when they are above green foliage. Herbicide is not applied to bare ground.

   Splatter guns may be used on either handlines or back packs to deliver a metered dose in spot spraying situations. As the name suggests they splatter the foliage and should be discharged from within, or over the plant or an area of soil.
4. Ultra low volume spraying equipment, such as controlled droplet applicators are becoming increasingly popular. This equipment produces very fine droplets of even size enabling effective coverage and hence control at very low rates of application.

   Because of the obvious cost savings, use of this equipment is becoming popular, particularly on small units such as carried by four wheel motor bikes and all terrain vehicles..
5. Ropewick Applicators. As have been previously discussed.

Calibration of equipment

There are two basic methods of spray application. The constant speed method where a rate per hectare is aimed at and the spot spraying method where a percentage spray solution is used and the plants are sprayed to the point of run off.

In any constant spray situation, whether it be with a boom or cluster jet, calibration is necessary. The aim of calibration is to find out the rate of fluid being delivered per hectare with the equipment moving at a particular speed. There are a number of methods for calibration of spraying equipment, the one outlined below is probably the most practical and accurate.

1. Check the nozzles and nozzle outlets to make sure they are all delivering the same amount of fluid. Any worn or defective nozzles should be replaced.
2. Measure out a distance of 100 metres.
3. Fill the spray tank to the top with plain water, making sure that spray lines are also full.
4. Adjust the pressure to the desired level and select the normal working gear and then spray the measured distance.
5. Measure the amount of water necessary to re-fill the spray tank.
6. Calculate output:

$$\text{Litres per hectare} = \frac{100 \times \text{No. of litres used on 100 metres}}{\text{Spray width in metres}}$$

7. The quantity of water used per hectare can be adjusted by varying the pressure or the speed.

Spray equipment should be checked and calibrated at the beginning of each spraying season and more often if possible.

Many modern boomsprays have computer monitoring which allow the operator to be more precise in herbicide application rates. This type of equipment has advantages in achieving the best possible results from the herbicide with cost savings by avoiding excessive rates of applications.

Efficient and safe spraying

1. READ THE LABEL and observe the safety precautions stated thereon.
2. Correctly calibrate equipment.
3. Use only clean water.
4. Do not spray on windy days.
5. Do not spray on frosty days.
6. Do not spray if rain is imminent.
7. Do not spray drought affected plants.
8. Spray on cool calm days when plants are making active growth.

Care and cleaning of equipment

Even the smallest traces of some chemicals, such as hormone herbicides, will cause harmful effects to other crops or corrosion may result from their residues. Thorough decontamination is necessary, particularly where the machine is being used for spraying susceptible crops. The following steps can be taken to reduce the possibility of harmful residues remaining in the unit.

* Water-soluble formulations
    1. Drain all liquid and then flush thoroughly with clean water.
    2. Use hot water containing household ammonia at 1:100, or washing soda at the rate of 10 ml to 1 litre of water. Rinse the tank thoroughly and flush through allowing the solution to stand in the system for

a minimum of 18 hours. Allow two to three days if cold water is used.
- Oil-based ester formulations
  1. Drain the equipment of all liquid and then flush out thoroughly with clean water.
  2. Rinse the tank thoroughly with kerosene and flush through.
  3. Rinse the tank with warm water plus liberal quantities of detergent and a solution of 450–900 g of washing soda per 115 litres of water. Leave the solution in the tank and lines for about five minutes.
  4. Flush several times with water, preferably hot.
  5. Replace all leather, plastic or rubber working parts such as hoses, washers, valves and packing rings.
- General cleaning and care
  1. Flush equipment after use, with clean water at low pressure. Remove jets so that dirt etc. can be removed from the lines.
  2. Check all hoses for wear and deterioration. Replace as necessary.
  3. Remove jets and filters, wash and store carefully.
  4. Never leave spray material standing in the spray unit after use. Spray material may deteriorate and the unit may be damaged by corrosive compounds in the herbicide. Clean the inside and outside of the tank. Check for rust and corrosion.
- Check before re-using boom equipment
  1. Inspect the tank, connecting pipes and fittings including jets for wear, corrosion or damage and replace faulty parts.
  2. Check the pump for efficiency. Output must be sufficient to maintain spraying pressures. When wettable powders are used in the spray mix, the pump's capacity will need to be 50 per cent greater to ensure both adequate spraying pressures and sufficient by-pass capacity for tank agitation. The agitator line should come from the pressure side of the pump and this line should deliver the liquid to the bottom of the tank. If an agitator jet is attached to this line, the effectiveness is increased. effectiveness is increased.

     Agitation may also be obtained by mechanical (i.e. a paddle or propeller) or by use of another pump.
  3. Calibrate equipment before re-use, replace any faulty jets.

Precautions and safety equipment when using herbicides

Toxic herbicides

Herbicides have varying degrees of toxicity, but they should all be handled the same and strict attention should always be taken towards the safety of the operator and the environment.

There are three main ways in which chemicals can enter the body:
1. *Orally*. Drinking the concentrate or spray mix.
   Oral intake is usually quite accidental and therefore rarely occurs if herbicides are correctly labelled and stored away from children. During spray operations keep chemicals and measures out of the reach of children. Make sure hands are washed before eating.
2. *Inhalation*. The breathing in of misted spray.
   Inhalation is usually readily detected because most herbicides have a characteristic smell. Even if the herbicide is not considered particularly hazardous, every precaution should be taken to reduce to a minimum the amount of spray inhaled. This can be done by wearing a respirator fitted with a special agricultural chemical filter.
3. *Skin Absorption*. Allowing concentrate or spray mix to contact skin areas.
   According to some studies this is the most common way that spray operators are poisoned, and yet it is often the least worried about. To avoid skin contact:
   - for relatively non-toxic chemicals wear overalls which cover the body including legs and arms and a sou-wester type hat over the head
   - for toxic materials wear PVC clothing to cover the whole body
   - use PVC gloves when handling concentrates or spray mix, combined with plastic splash-proof visor
   - wash hands or any part of the skin immediately with soap and water if contaminated with herbicides
   - wash all clothing used during spraying operations, daily.

Protective equipment available

1. Respirator. A respirator should be worn when spraying toxic chemicals. Use the correct filter. Dust filters or fumigant filters do not afford protection when using the bulk of agricultural herbicides. It must be fitted with an agricultural chemical filter or cartridge.
   Important: Cartridge life is limited and can be as little as two hours in extreme conditions, and about two weeks under normal conditions. Once the cartridge has become neutralised by chemicals it is useless and unless changed could present a greater hazard than the use of no respirator at all.
2. Face shield. A clear plastic face shield is recommended when handling or mixing liquid concentrates to protect the face against splashes.
3. PVC Gloves. Impervious gloves used when handling or mixing chemicals. Do not use canvas or other gloves which absorb chemicals, and thus increase the danger of chemical contact with the skin.
4. Overalls. Ordinary long sleeved overalls provide adequate protection

for most spraying jobs. They should be washed daily and laundered separately from other clothes.
5. PVC Jackets and Trousers. These are worn in place of overalls where skin absorption is a real danger.
6. Hat. A canvas type sou-wester hat is suitable protection for most herbicides. Where further protection is required a PVC sou-wester is available.
7. Rubber Boots. These are useful if it is necessary to walk through freshly sprayed crops.
8. Chemical goggles. Provide eye protection to skin absorbed chemicals and prevent irritation from other chemicals.
9. Soap and water. Should be readily available at all times to wash hands and face before eating or smoking or to wash any part of the body contaminated with the spray.

Deciding what protective equipment to use

The label: Read it, Heed it! The safety directions on the label give an exact description of the minimum precautions to be taken. If only there was some sort of guarantee that all users would follow these directions then there would be no problem and no need to emphasise the precautions as set out in this section.

If poisoning is suspected seek medical advice immediately — making available to the doctor information on the type of chemical or chemicals suspected.

If possible take the chemical container to the doctor or if this is impractical make sure you inform the doctor of both the trade name and active ingredients of the herbicide.

The active ingredient or active constituent is printed on the label. For example:

Trade name	*Active constituent*
Metham Total Soil Fumigator	400 g/L metham-sodium salt

Doctors may be unfamiliar with many of the agricultural chemicals used. You can however obtain complete information on any chemical by phoning the closest Poisons Information centre. Material Data Safety Sheets are also available from chemical manufacturers.

7
Fertiliser requirements and grazing management

Plant nutrients

Sixteen different elements are required for plant nutrition to promote growth and reproduction. All except carbon, oxygen and hydrogen come from the soil. Carbon and oxygen are obtained from air and hydrogen from water. Nutrients required in large amounts are:

Element	Chemical Symbol
Carbon	C
Hydrogen	H
Oxygen	O
Nitrogen	N
Phosphorus	P
Potassium	K
Calcium	Ca
Sulphur	S
Magnesium	Mg

Nutrients required in smaller amounts include:

Element	Chemical Symbol
Iron	Fe
Copper	Cu
Zinc	Zn
Manganese	Mn
Boron	B
Molybdenum	Mo
Cobalt	Co

Some plants also require Sodium (Na) and Chlorine (Cl), while stock may also need Iodine (I), Fluorine (F) and Selenium (Se).

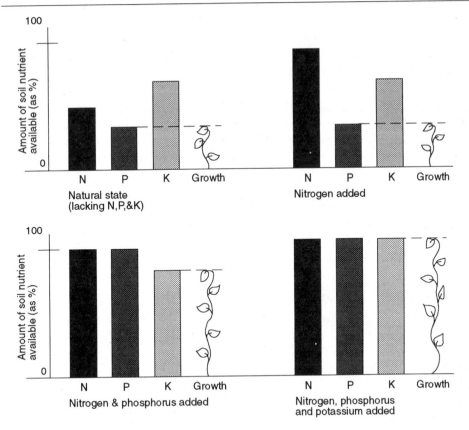

Figure 7.1 The most deficient nutrient limits plant growth

The elements required by plants in large amounts are known as major elements. Those needed in only small amounts are minor or trace elements.

The nutrient elements are stored in the soil in clay mineral fragments and organic material.

All nutrients must be available in adequate amounts for plant growth. If any are deficient growth will be restricted by the deficiency — even though there may be an abundance of all the other required elements. This phenomenon is known as the Law of Limiting Factors and is clearly illustrated in Figure 7.1 above.

Law of limiting factors

To promote maximum pasture growth it is therefore necessary to establish and maintain a balance of nutrients by using a range of soil fertility managment strategies to correct deficiencies and to replace nutrients removed in produce.

Just as inadequate fertilising may result in deficiencies, excessive amounts of some elements can cause problems by making others unavailable, e.g.

Pasture management

excessive molybdenum reduces the availability of copper and excessive calcium ties up zinc, iron and manganese.

Some deficiencies are therefore caused by either the lack of a particular nutrient or by its unavailability.

The presence and availability of particular nutrients can be determined by soil tests and tissue analysis. Paddock history and a knowledge of nutrient removal also provide a useful guide to requirements as do nutrient deficiency symptoms in pastures. It is also possible to determine the nutrient status of a crop by its degree of vigour or by obvious signs of deficiency.

Nutrient removal

Removal of nutrients is a continuous process. Losses are caused by:
- soil erosion
- leaching
- volatalisation
- produce taken from the land.

A knowledge of quantities removed by grazing, hay production or cropping provides a basis for planning nutrient management programs.

Table 7.1 indicates the losses of elements that can be expected under normal conditions of soil, climate and management.

Table 7.1 Nutrient loss by production

Produce	Yield	Loss in kg			
		N	P	K	S
Meat	50 kg live wt.		0.4		0.4
Milk	1000 litres		1		0.6
Wool	5 kg greasy		0.02		0.2
Pasture hay	1 tonne	50	9	50	
Lucerne hay	1 tonne	27	3	16	2.5
Cereal hay	1 tonne	20	2	18	1.4
Wheat	1 tonne grain, 1.6 tonnes straw	28	5	20	3
Barley	1 tonne grain, 1.4 tonnes straw	18	5	20	3
Oats	1 tonne grain, 1.7 tonnes straw	28	6	30	4
Peas	1 tonne grain, 1.6 tonnes straw	60	7	30	4

1 kg N is equivalent to	2 kg Urea	1 kg K is equivalent to	2 kg Muriate of Potash
	5 kg Sulphate of Ammonia		2.4 kg Sulphate of Potash
	3 kg Ammonium Nitrate	1 kg S is equivalent to	6 kg Gypsum
1 kg P is equivalent to	12 kg Superphosphate		9 kg Superphosphate
	5 kg Triple super		67 kg Triple super

Deficiencies

Deficiencies in pastures are often difficult to recognise and interpret. Many symptoms are very similar and confusing. This situation may be aggravated by adverse soil conditions.

Soil conditions that cause deficiencies are:

- Lack of nutrients — in soils that are either naturally deficient or soils in which a deficiency has been induced by failure to replace nutrients that have been removed in produce.
- Incorrect pH — extremes of pH drastically affect the availability of nutrients.
- Excessive amounts of organic matter — these tie up some elements and make them unavailable. Organic matter can also provide a constant supply of nutrients.
- Nutrient imbalance — a deficiency or over supply of one element may affect plant growth despite the presence of all other elements in the correct quantities.
- Poor aeration — causes soil micro-organisms to obtain oxygen for respiration from other substances in the soil. This reduces their activity and ability to break down organic matter and other compounds to release nutrients.
- Extremes of temperature — high and low soil temperatures reduce the availability of nitrogen and reduce the activity of soil micro-organisms.
- Soil moisture — very low moisture levels reduce nutrient availability and waterlogging causes poorly aerated conditions also reducing nutrient availability.

Deficiency symptoms (Tables 7.2 and 7.3) in Australian pastures are caused mainly by lack of nitrogen, phosphorus or potassium but other deficiencies do occur and produce symptoms in specific situations. Molybdenum deficiency is one such deficiency that is common in legumes on acid soils.

Deficiency symptoms are a useful guide but should be confirmed by soil tests and plant tissue analysis. Consult your local Department of Agriculture regional advisory staff for confirmation based on local knowledge before finalising your nutrient management program, especially when trace element deficiencies are suspected.

Table 7.2 Symptoms of nutrient deficiencies

Element	Symptoms	Cause and Correction
Nitrogen (N)	Leaves turn pale green then yellow and die back from the tips. Plants do not thrive and become stunted.	Possible causes are cold waterlogged soils, excessive soil organic matter or poor legume nodulation. Apply N in suitable climatic/soil conditions. Co and Mo may be required for nodulation.
Phosphorus (P)	Seedlings establish poorly. Leaf and root growth is stunted. Leaves dull blue-green and tend to turn purple in in cold weather.	Naturally deficient in most Australian soils. Correct by supplying adequate superphosphate seeding and top-dressing in the autumn.
Potassium (K)	Stunted growth with leaves dying back from the tips and margins. Clover leaves turn reddish brown, grass leaves yellow between the veins. Susceptibility to disease and cold increases.	Commonly induced in hay paddocks. Apply K in late winter or early spring.
Calcium (Ca)	New growth is stunted and distorted. Leaf margins and interveinal areas turn reddish brown.	Apply lime if pH is less than 6 to raise pH to 6.5. Lime pellet legume seed.
Sulphur (S)	General yellowing of plants. Foliage develops a papery texture.	Naturally deficient in most Australian soils. Apply single superphosphate, gypsum or sulphur fortified super-phosphate.
Magnesium (Mg)	Yellowish stripes that turn to red appear on the foliage, particularly the older leaves.	Deficiencies seldom encountered except in vegetables such as tomatoes.
Iron (Fe)	Young leaves turn yellow between the veins. There is no initial stunting, but as older leaves become affected new new growth is stunted.	Caused by excessive lime in the soil. Apply iron sulphate spray to correct.
Copper (Cu)	Entire plants become stunted and yellow. The growing points and tips of younger leaves die. The edges of clover leaves turn white, flowering and seed setting are reduced. Symptoms are often first noticed in livestock — deficiency causes steely wool, and loss of pigmentation in the coats of cattle and wool of black sheep.	Sometimes occurs naturally or is induced by a lowering of the pH. Apply copper at seeding, and where acute deficiencies occur such as on peat soils, top-dress annually.
Zinc (Zn)	Growth is stunted, leaves become thin, shrivelled and dark in colour. The leaves appear dry.	Low temperatures may induce deficiency. Apply zinc in super-phosphate.
Manganese (Mn)	Leaves turn pale green between the veins and look striped or spotted.	Caused by high levels of lime or organic matter. Cereal crops are particularly susceptible. Apply manganese sulphate at seeding and follow up with foliar sprays of the same material every four weeks if necessary.
Boron (B)	Growth is stunted and dieback occurs at the growing points. Yellow streaks appear in the leaves.	Deficiency is rare, but to correct apply boron in superphosphate.
Molybdenum (Mo)	Similar symptoms to those caused by lack of nitrogen.	May be induced by increasing acidity. Apply molybdenum in superphosphate at seeding or when topdressing.
Cobalt (Co)	Plant symptoms are rare. Livestock, especially young stock fail to thrive and waste away. Wool production falls.	Occurs naturally in some areas. Apply cobalt in superphosphate at seeding and when topdressing every 3–5 years. Drench livestock or treat with cobalt bullets.

Table 7.3 Soils likely to be deficient

Nutrient	Soil Type
Nitrogen (N)	Sandy soils, and soils in high rainfall areas with low or very highlevels of matter. Soils with poor legume history.
Phosphorus (P)	Most Australian soils, particularly those that are highly leached.
Potassium (K)	Sandy soils, organic soils, heavily cropped soils, leached and eroded soils and acid soils.
Calcium (Ca)	High rainfall sandy, acid and clay soils.
Sulphur (S)	Soils containing low levels of organic matter.
Magnesium (Mg)	Acid, sandy and organic soils often associated with calcium deficiency.
Iron (Fe)	Poorly drained soils. Soils with high levels of calcium, phosphorus, magnesium and copper.
Copper (Cu)	Organic soils, alkaline soils, soils with a high calcium level, leached sandy soils and soils with a high nitrogen level.
Zinc (Zn)	Highly acid soils, leached sandy soils, highly alkaline soils and often ironstone soils.
Manganese (Mn)	Highly alkaline soils, limy soils, light siliceous sandy soils and peats. Can be at toxic levels in acid soils.
Boron (B)	Sandy acid leached soils, alkaline soils with free calcium.
Molybdenum (Mo)	Ironstone soils; and well drained acid soils.
Cobalt (Co)	Alkaline soils low in clay and humus. Usually associated with copper deficiencies.

Soil tests

Soil tests are a useful basis for making decisions on soil nutrient management programs because they reduce the guesswork involved in soil management and fertiliser use. They provide guidelines on the type and amount of fertiliser to apply and they also indicate situations where the use of trace elements and the use of lime should be considered. Soil tests are also useful in determining the most suitable species to plant. In established pastures the regular monitoring of key indicators (phosphorus, potassium and pH) is one of the more practical uses of soil tests.

Soil pH

Soil pH is the measure of its acidity or alkalinity. It is very important because many pasture and crop varieties prefer certain ranges of pH.

Lucerne, annual medics	6.5–8.5
Subterranean clover	5.5–7.0
Phalaris, wheat, barley	6.0–8.5
Lupins	5.0–7.0

Pasture management

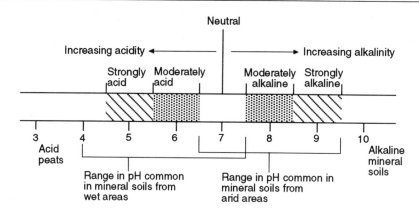

Figure 7.2 The pH scale

Soil pH can be measured using either the 1 part soil: 5 parts water method or the calcium chloride method. Before attempting to interpret results of a soil pH test it is necessary to know which method was used.

Table 7.4 The effect of pH on the availability of plant nutrients

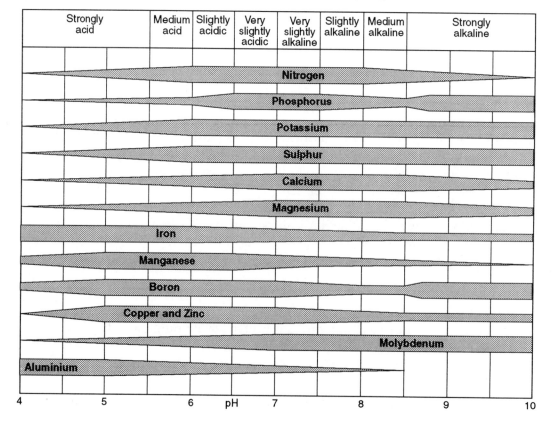

Extremes of pH should be corrected. Very acid soils can be improved by the addition of lime. The approximate quantity of ground limestone required to reduce the acidity in the top 20 cm of soil and to raise the pH to 6.5 depends mainly on the soil type and the initial pH level, as shown in Table 7.5. Note this table is only a guide and soil tests should be conducted to determine lime requirements.

Table 7.5 Lime required to raise pH to 6.5

Initial pH	4.5	5.0	5.5
Sand, sandy loam	2.5 tonnes/ha	2.0 tonnes/ha	1.0 tonnes/ha
Loam	7.5	5.5	3.5
Clay loam	10.0	8.0	6.0

The effectiveness of the lime will be improved if it is incorporated into the soil rather than broadcast onto the surface of the soil.

Organic carbon

This test is generally used as a guide to the fertility of soils.
The test categorises organic carbon levels as:

Low	< 1.0%
Normal	1.0–2.0%
High	> 2.0%

When low results are obtained you should try to build up the organic matter content of the soil. Pastures are a good method of building up soil organic matter levels.

Potassium test

Potassium levels are obtained by determining the potassium concentration in the sodium bicarbonate soil extract test (also used to determine soil phosphorus levels).

Deficiencies are most likely to occur in high rainfall areas, deep sands or sand-over-clay soils, especially in paddocks regularly cut for hay.

The test is interpreted as follows:

Soil K Test		*Suggested fertiliser*
Very low	< 50 mg/kg	Apply 25–50 kg K/ha for grazing pastures.
		50–100 kg K/ha for hay paddocks.
Low	50–80 mg/kg	50–100 kg K/ha for hay paddocks.
Marginal	80–120 mg/kg	25–50 kg K/ha when hay is cut continuously.
Adequate	> 120 mg/kg	No K necessary.

Potassium applications are most effective in late winter or early spring.

Table 7.5 General interpretation of soil tests for sodium bicarbonate extractable P and general guidelines for fertiliser use. (Rainfall < 800 mm.)

Soil test classification*	Soil P test (mg/kg)		Fertiliser guidelines	
	Sands	Other soils	Warm temperate zone soils	High rainfall pasture soils
Very low	< 10	< 15	Apply up to 20 kg P/ha. Lower application rates of P are required for the < 300 mm rainfall areas. Drill fertiliser with seed. Pastures should be topdressed; if not, higher application rates of P should be drilled with crop.	Sands: Apply sufficient P to meet pasture annual requirements. In leaching situations delayed applications are suggested. Rate: 10 to 15 kg P/ha depending on rainfall and stocking rate. Other soils: 5 to 10 kg P/ha will maintain pasture production. Higher rates will improve soil P reserves.
Low	10–20	15–25	As for very low classification.	Sands: Similar to low classification, except that application rates of P can be reduced slightly. Other soils: Similar to low classification, except that application rates of P can be reduced slightly.
Marginal	20–25	25–30	Increase current application of P slightly. Drill at least a small quantity of P fertiliser with crops. Pastures should be topdressed; if not, higher application rates of P should be drilled with crop.	All soils: Maintenance application rates of P are required (5 to 10 kg P/ha).
Adequate	> 25	> 30	Maintenance application rates of P are required (3 kg P/tonne of cereal grain harvested).	All soils: Maintenance application rates of P are required (5 to 10 kg P/ha). Pasture production will not suffer from P deficiency if fertiliser is omitted for a season.

* No soil test values for P have been defined for the highly calcareous and the ironstone soils. Fertilise these soils for current crop needs.

Phosphorus test

The sodium bicarbonate soil phosphorus test provides a reliable estimate of soil phosphorus available for crop and pasture production. It should be used to formulate practical fertiliser application programs by increasing or decreasing application rates or by maintaining current rates depending on existing levels of phosphorus. Another widely used phosphorus test is the Bray test. It is important that the method of soil phosphorus testing is known before making any recommendations. The level of rainfall in a district will also affect the interpretation of the soil phosphorus test. It is advisable to consult an agronomist for soil test interpretations.

The amount of phosphatic fertiliser to apply can be calculated by reading Figure 7.3.

Remember that Double or Triple Super does not contain as much sulphur as single superphosphate. Continuous use of either will result in deficiencies of sulphur, particularly likely on light sandy soils under crop.

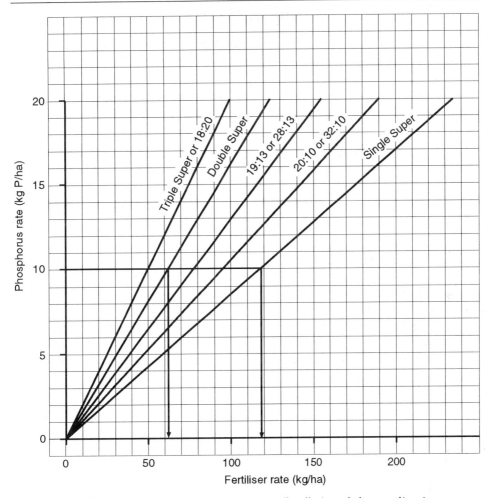

Figure 7.3 Relationship between P rate (kg/ha) and the application rate of various P fertilisers

Other soil tests

Soil nitrogen is measured only in the nitrate form. It is difficult to get an interpretable result from this test since the level of nitrates in the sample can change quite dramatically after sampling. Refrigeration of soil samples helps to prevent this.

Soil salinity is assessed by measuring the total soluble salts in the soil, and by measuring the soil's electrical conductivity.

Exchangeable cations are measured to assess the need for lime in acid soils, gypsum in saline soils and gypsum in poody structured soils.

It is difficult to correlate the soil tested levels of many trace elements with plant growth. For this reason it is uncommon to test for trace elements in the soil. Plant tissue tests are commonly used to determine trace element deficiencies and toxicities.

Soil sampling

The value of soil tests depends largely on the samples tested. Unless they are truly representative of the area to be fertilised then the test results will be misleading.

The greatest source of error in soil test results is the sampling procedure. Samples must be taken to a full sampling depth of 10 cm in pastures. This is because there is a large change in soil nutrient levels over a relatively small depth in pasture soils. As fertilisers, such as superphosphate, are applied they have a higher concentration near the surface and this drops rapidly with soil depth. Uneven depth of sampling will result in a biased soil test result. If samples have been too shallow the resulting soil nutrient levels will be higher than if samples had been taken at a constant and correct depth.

Avoid unusual areas such as timber burns, headlands, sheep camps or swampy areas. Including soil cores from these areas will not allow a representative sample to be obtained. If these areas have to be tested they should be tested separately.

Different soils should be tested separately. If cores from different soils are mixed, the result of the soil test will not be indicative of the fertility status of either soil. Several samples should be collected in each paddock to check the variability that exists in the paddock's fertility status.

For each soil sample there should be at least 12 cores taken. This is necessary to ensure that there are enough cores to get a representative sample for testing. More than 12 cores may be taken for a sample, however if too many are taken problems occur with processing and sub-sampling the soil sample.

Soil testing is undertaken by some state Departments of Agriculture, or by private laboratories and fertiliser companies. Consult your local agronomist for the most accessible.

Plant analysis tests

Plant tissue analyses are used commercially for diagnosing nutrient deficiencies, toxicities or mineral imbalances. This data can be used for predicting nutrient deficiencies in current or future crops and for making fertiliser recommendations. Like soil tests they can also be used for monitoring the effectiveness of current soil fertility practices.

Plant tissue testing also has many limitations in that the results can easily be misinterpreted. Professional advice should be sought for interpretation of tissue analysis.

There are a number of commercial tissue tests available. One commonly used in pasture situations is a leaf analysis for the phosphorus content of subterranean clover pastures. This is available through some fertiliser companies.

Fertilisers

There are many fertilisers — organic and artificial (chemically manufactured). To choose the best for a particular situation you need to know the nutrient elements required and:
- The analysis of the range of fertilisers that contain the required elements in different proportions.
- The form that elements are in — whether or not they are available to plants.
- The cost per unit of nutrient element.
- The suitability of the fertiliser for either drilling, broadcasting or applying the fertilisers.
- Other benefits that the fertilisers may provide (such as improving soil structure with organic fertilisers such as chicken litter).

Make sure that you familiarise yourself with these details before making your choice.

Application methods

Fertilisers are applied to crops and pasture by:
- Drilling
- Topdressing
- Irrigation
- Foliar sprays.

Drilling in with the seed enables the most accurate distribution and placement of the fertiliser. It can be placed at the required rate in contact with the seed, or in bands within the anticipated root zone, below or beside the seed.

Topdressing is the most common method of fertilising established pastures. Although not as accurate as putting the fertiliser out through a drill or a dropper, broadcasting, from aircraft or ground vehicles is a relatively quick and practical method. Topdressing with superphosphate is common in pasture situations, particularly with established pastures.

To avoid missing strips when broadcasting it is wise to halve the application rate and treat the area twice.

Irrigation provides an excellent method of applying soluble fertilisers (such as urea). The urea is dissolved in a tank connected by a venturi to the irrigation line and an even spray pattern from the sprinklers ensures even distribution of the fertiliser.

Foliar sprays from aircraft or ground driven boomsprays enable high analysis liquid fertilisation for a quick response.

Fertiliser programs

When formulating fertiliser programs for pastures remember to:
- Determine the nutrient requirements first.
- Correct soil conditions that may induce deficiencies.
- Correct minor or trace element deficiencies.
- Topdress early enough to ensure germinating annual pasture plants do not suffer from a lack of P.
- Apply K to hay paddocks in late winter or early spring.
- Apply N to grass dominant pastures prior to rain in autumn or spring when a quick boost is required, or when sowing areas with a poor history of legumes.
- Keep detailed, accurate records of fertilisers applied and production.

Grazing management

Grazing management is an art that must be learnt from experience.

It should aim to obtain economic returns from pastures by maximising grazing value and minimising losses of surplus feed. Management should also encourage pastures to persist and resist invasion by weeds.

Management practices should be influenced by:
- age and type of pasture
- pasture varieties
- seasonal conditions
- type of livestock
- soil types
- topography
- economics.

Newly sown and renovated pastures

Good management during the year following sowing is essential if new pastures are going to persist or regenerate satisfactorily.

Initially, new pastures should not be grazed until they are 15 cm high. They should then be grazed quickly, using as many stock as possible, to a height of not less than 5 cm and allowed to recover. Repeat the process as growth dictates. In this way all pasture varieties are encouraged to develop evenly and thicken. None will dominate and smother other less vigorous or less advanced plants.

This is particularly important for new plants that have been sod seeded amongst established perennials.

If sufficient stock are not available to graze the pasture quickly grazing is likely to be selective and the more palatable varieties will suffer. In this case it is wise to mow the pasture to the desired height.

Sheep are best for grazing new pastures. They are less likely to pull up plants or pug them into the ground (causing permanent damage) than cattle or horses. This is especially so on light or very wet country.

Following the initial grazings, as plants thicken and develop stronger root systems, grazing pressure should gradually increase. Check the new pasture regularly for insect damage and excessive weed growth. Both should be controlled promptly if necessary.

New pastures should not be overgrazed or cut for hay during the first year. This allows annuals to set sufficient seed for satisfactory regeneration the following year and encourages perennials to develop a deeper root system making them more drought tolerant and persistant.

Established pastures

Ideally, grazing management for established pastures should be similar to that described for newly established or sod seeded pastures. This will assist in the longevity of the pasture and optimal livestock production from the pasture. Grazing pressure must be regulated according to seasonal growth to gain maximum benefit from pastures.

Undergrazing leads to selective grazing. Less palatable varieties become tall and rank, then flower, set seed and proliferate. The more palatable varieties are constantly grazed and disappear.

Overgrazing weakens pasture plants and may prevent seeding. This leads to plant mortality or lack of regeneration and bare patches develop. These are quickly filled by weeds.

The grazing capacity of pastures is constantly changing — reducing during summer, to early autumn, low in winter and high in spring. As grazing demand tends to be more constant, grazing pressure is usually a compromise between what is best for the livestock and what is best for the pastures. Most graziers favour their livestock when they should favour the pastures.

However, apart from providing adequate fertiliser each year and effectively controlling pests and weeds there are a number of management factors and grazing practices that can be used to get the best out of available pastures.

Management factors

- Paddock design — paddocks should suit the topography and property enterprise to enable even grazing throughout, e.g. problem areas such as those that are prone to erosion or waterlogging should be separately fenced.
- Watering facilities — these must be within easy reach for stock from all quarters of the paddock to encourage even utilisation.
- Shade trees — should be evenly distributed.

Grazing systems

There are a number of different grazing practices for different situations:

- Continuous grazing. This system leaves stock in one paddock for a year or more. It is widely practised in the pastoral zone. Stock are mustered only for shearing, crutching, marking and sale. Stocking rates are critical in this fragile environment and have to be constantly reviewed to prevent the native vegetation from being eaten out in times of drought.
- Set stocking. Under this system livestock are not left in the same paddock for as long. Practised in the high rainfall districts, the system aims to give the best feed to the stock that require it most (e.g. lambs) while minimising stock movement. Ewes are usually left in one paddock from the time of mating until lambing, or from lambing until weaning.

    Lambing percentages are invariably high in set stocked flocks and young stock thrive better on the constant plane of nutrition with few disturbances in familiar surrounds. Labour requirements are also minimised with this system.
- Rotational grazing. This system involves intensively grazing a paddock for a short period of time (up to one week), moving the stock to a fresh paddock and allowing the original paddock to regrow for up to five weeks.

    This system is particularly effective for dryland lucerne pastures and mixed irrigated pastures. They persist and produce well under this strictly regulated grazing pressure.
- Time controlled grazing. Also called cell grazing this is another strategy that is similar to rotational grazing. Cell grazing aims to manage pastures so that they have adequate time periods to recover from grazing and optimise productivity by aiming at keeping pasture in its optimal stage of production for as long as possible. This form of grazing management has only just been introduced to Australia but it has proven successful in other countries.
- Strip grazing is a modified form of rotational grazing that utilises electric fences in front of and behind the stock (usually dairy cattle) to give them a fresh portion of a pasture (usually daily) while allowing other strips to recover.
- Deferred grazing. This practice involves holding stock in small paddocks and hand feeding them for about six weeks following the opening rains. This enables seedling pastures to get away more quickly than they would if grazed from the break.

Pasture production in both quantity and quality vary considerably throughout a year. This variation will result in different levels of production from the livestock grazing the pastures. In the past it was common practice to estimate the carrying capacity of a pasture on a yearly basis but the recent trend in grazing management is to continually assess the quantity and quality of pasture production and to estimate the level of production that is likely from the various categories of livestock grazing the pasture.

To achieve this the grazier must first be able to accurately assess pasture quantity and quality. This initially requires sampling, drying, weighing and partitioning pastures samples. With experience graziers become reasonably accurate at estimating pasture quantity and quality.

The results from the assessment process can be interpreted using a computer program developed by the CSIRO called "Grazfeed". The Grazfeed program will determine the level of livestock production likely to be achieved from the pasture and more importantly identify the limiting nutritional factors that the pasture is not capable of supplying to the particular category of livestock. In New South Wales this system of pasture assessment is called "Pro-Graze".

Fodder crops and conservation

In southern Australia about 75 per cent of pasture growth is produced in the spring; far more than can be utilised by grazing unless stocking rates are dramatically increased during this period. Unfortunately standing pasture rapidly declines in quality as it dries off in summer and losses due to trampling and decomposition are likely to exceed 50 per cent.

This waste can be prevented by conserving the excess spring pasture growth as either hay or silage, both of which can be used for either drought reserves or feed supplements.

Supplementary feeding can be carried out at either maintenance or production levels. Maintenance feeding will allow livestock to maintain weight and condition but has no economic gain to the producer. Maintenance feeding is usually carried out under conditions of pasture shortage in either quantity and quality and is aimed at carrying the animals through to periods where pasture conditions improve. Maintenance feeding is most common with pregnant and lactating animals.

Production feeding aims at keeping animals on full levels of production. This is usually aimed at livestock that are being grown out or fattened for market. This type of feeding aims at producing greater economic returns over the cost of the feed.

Apart from conserving fodder, land utilisation can often be improved by growing short term fodder crops for either grazing, green chop, silage or hay. Crops such as oats or barley sown in the autumn or millet, hybrid forage sorghums, hybrid maize or the brassica crops such as rape, turnips or chou moellier sown in the spring can greatly improve the short term productivity of land, and can also often be used in a pasture renovation program to assist in controlling weeds prior to resowing permanent pasture. These spring sown crops are particularly valuable where irrigation is available.

Fodder crops for summer and autumn feed

Brassica crops

Brassica crops can be spring sown between September and November under Australian conditions. Their main use is where lack of labour and irrigation prevent other fodder crops from being grown. Usually they are grown under dryland conditions but they can be irrigated.

Soil requirements: All do best on well-drained, loamy soils of good fertility, although turnips can be grown on relatively infertile sands, and mangels on heavy salt-affected ground.

Seedbed requirements: Seedbed preparation should be aimed at preparing a fine, friable, weed-free area early in winter, so that the crops can be sown in early spring.

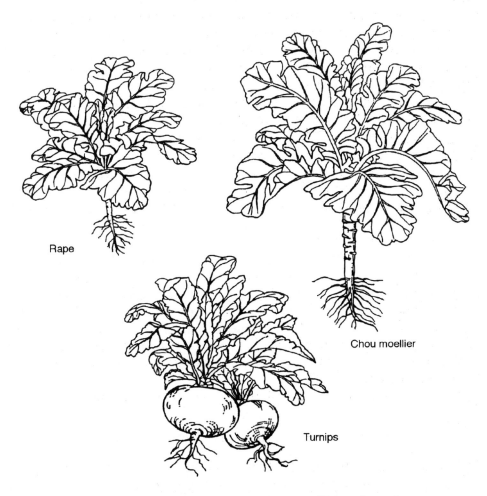

Figure 8.1 Brassica crops; rape, chou moellier and turnips

Seeding: Low rates should be used in all but late districts or where irrigation is available, and the seed should be sown either through a small seeds-box or mixed with reverted superphosphate, broadcast and lightly harrowed. If broadcasting, the time that the seed and the fertiliser is mixed must be kept to an absolute minimum, otherwise the germination rate will be adversely affected. The seed should be sown 1–2 cm deep.

Fertiliser requirements: Superphosphate should be applied at 120 kg/ha when sowing the crop in areas with a good super history, but this rate should be increased to 200 kg/ha in relatively infertile paddocks.

Pest Control: All these crops are attacked by aphids, cabbage centre grubs, cabbage white butterflies, and various cutworm. A light grazing will often control aphids, cabbage moths and butterflies, but insecticides may be required to control severe infestations and some of the other pests.

Grazing management: Brassica crops taint milk. Dairy cattle should not be allowed to graze these crops for two hours prior to milking. Strip-grazing is the most efficient way to use fodder crops, minimising losses through trampling and fouling and enabling maximum recovery following grazing.

Rape will mature in ten weeks to provide excellent grazing for maintaining milk production or for fattening sheep and lambs in early summer when other pastures dry off. The crop should be grazed as the plants mature and the leaves begin to turn purple in colour. Following grazing recovery of the crop the availability of further grazing will depend on summer rains or irrigation. It is not very drought resistant.

Rape varieties include Rangi and Moana which have some resistance to aphids, Dwarf Essex and Giant Emerald.

Turnip tops can be grazed lightly during summer providing they are growing well and not suffering from moisture stress, and then the crop can be heavily grazed when the bulbs mature in autumn. Yields in excess of 30 tonnes of turnip bulbs per hectare can be grown under good conditions. Grazing should be finished before winter to prevent losses due to the bulbs rotting.

Mammoth White Purple Top turnip is the variety usually grown.

Chou moellier, whether sown alone or as a cover crop for spring-sown perennial pastures such as phalaris, will establish and hang on during the summer and come away quickly after the autumn break, providing a large bulk of feed for sheep or cattle. Grazing can start when the plants are established well enough not to be pulled out by the stock, and regulated to prevent more than the top third of the stalk being stripped of leaves, until after the autumn break.

Mangels and swedes require similar management to turnips.

Millet

Millet is a summer growing grass crop suitable for grazing with sheep or cattle, or for making hay or silage. It can be grown to supplement irrigated pasture, or used as part of a pasture renovation program.

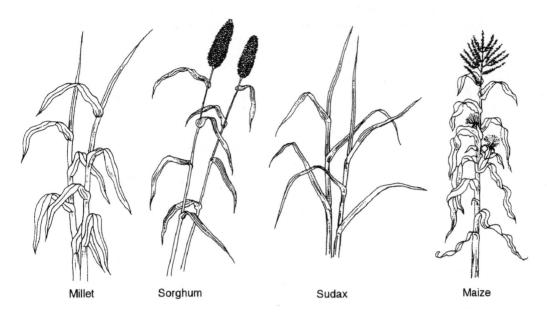

Figure 8.2 Millet, sorghum, sudax and maize

There are no prussic acid poisoning risks with millet as there are with the hybrid forage sorghums and it is safe to graze at all stages of growth. Two varieties are Japanese and Shirohie, a later-maturing variety.

Soil requirements: Millet produces best on well-drained fertile loams.

Seeding requirements: Sow into a well-prepared weed-free seedbed after mid-September.

Seeding: Sow 1–2 cm deep at a rate of 20–30 kg/ha under irrigation or 7–12 kg/ha under dryland conditions.

Fertiliser requirements: Sow with 200–300 kg/ha superphosphate and sulphate of ammonia 1:1 or 2:1 and topdress with 50–100 kg/ha urea after each grazing.

Grazing: Strip graze when the crop is 15–30 cm high. Graze out once it runs to head.

If made into hay, millet should be put through a conditioner.

Hybrid maize

Hybrid maize can be grown either under irrigation or dryland conditions in districts where the rainfall is at least 625 mm (25") per annum. Planted during November, December and January, maize will grow to maturity in 16–20 weeks to provide a huge bulk of feed in the summer, autumn, early winter period that can either be strip grazed with cattle, fed as greenchop or ensiled. Greenchop yields in excess of 30 tomes per hectare are common.

Soil requirements: Maize does best on well-drained fertile loams.

Seedbed requirements: A clean weed-free seedbed is essential.

Seeding: Seeding cannot be carried out until ground temperatures measured in the middle of the day reach 16°C at a depth of 10 cm, otherwise germination will not be satisfactory. The seeding rate will depend upon the grade of seed and whether or not the crop is to be irrigated. It should be sown 2–3 cm deep, 21 cm apart in approximately 75 cm rows. Seeding rate and placement are critical and extreme care should be taken to ensure accurate and even distribution of the seed at the rate recommended by the seed producer. This information is always available on the sack or from your seed merchant.

Fertiliser requirements: Sow with 200 kg/ha of 1:1 or 300 kg/ha of 2:1 superphosphate and sulphate of ammonia, and then apply additional nitrogen in the form of urea or sulphate of ammonia as side dressing, prior to watering, when the crop is 30 cm high.

Irrigation: The crop lends itself to furrow irrigation, and sufficient water is required to be able to supply at least 400 mm (16") at the rate of 25 mm (1") per week throughout the growing season to ensure high yields. Because maize grows about three metres tall sufficient space must be left in the crop to move sprinklers with risers or for travelling irrigators if these methods of watering are to be used.

Harvesting and Grazing: Maximum benefit will be obtained by harvesting or grazing this high energy crop when the grain on the cob is at the soft dough to denting stage. Once used the crop will not regrow, but once mature it can be left standing until required without any great loss of yield or nutritive value.

Hybrid forage sorghum

There are numerous hybrid forage sorghum varieties available including Sudax ST6, Magic, Honeydrip and Jumbo.

Like millet they can all be used to supplement irrigated pastures during summer and autumn, used as part of a pasture renovation program, or made into silage or hay. However, if made into hay, conditioning is essential, to ensure quick and even drying.

The hybrid forage sorghums have the ability to outyield millet during the hot summer months provided growth is not limited by either lack of nutrients or lack of water.

All these crops have one major disadvantage when compared to millet. They all contain levels of prussic acid which can be toxic at certain stages of growth, which can make grazing risky unless it is strictly controlled.

They should not be grazed until they are at least 45 cm high, or immediately following a growth check caused by either moisture stress, frost or insect attack.

Growing and management requirements for these crops are similar to those of millet but they cannot be sown as early as millet, requiring a soil temperature of at least 16°C for satisfactory germination. Planting of these crops should be delayed until mid-November.

Fodder crops for winter feed

Winter production is generally maintained by the following practices:
- Pasture grazing
- Feeding bought fodder and concentrates
- Feeding of excess summer and winter fodder and pastures previously conserved either as hay or silage on the farm
- Access to winter fodder crops either grazed or foraged.

The following crops may be considered and worked into the overall plan to supply winter feed.

Oats

The main winter fodder crop is oats, which may be sown over a period extending from February to May. Oats may be used for grazing, green feed, silage or hay — or it may be grazed for a time and allowed to mature for hay.

Oats is either sown into a prepared seedbed, or sod seeded into existing pastures. For best results from oats in a prepared seedbed, sow on a well-prepared seedbed preferably placing the seed and fertiliser into a moist firm seedbed with a drill.

Sow February to May at 70–120 kg seed/ha. Fertilise with superphosphate and sulphate of ammonia 3:1 at the rate of 100–300 kg/ha. A further application of nitrogen may be necessary following the first grazing.

Oat varieties include those for early grazing, late grazing, and those for grazing and hay and should be selected accordingly.

Field peas or vetch may be included with the oats to provide additional protein. In this situation sow the oats at 30–35 kg/ha, with field peas at 70 kg/ha, or vetch at 25–30 kg/ha.

The sod seeding of oats into existing pastures to supply additional grazing during the winter is a common practice.

A clover/ryegrass pasture sod-seeded with oats will provide forage needs from May to September, particularly where irrigation is available.

Time of seeding is an important factor — late February–early April — when the mean daily temperature is approaching 18°C, will ensure rapid germination.

The oats should be sod seeded into moist pasture after hard grazing or mowing. If needed, irrigate before sowing. Where summer growing grasses like paspalum are very vigorous, sowing should be delayed until early April.

Sow 120–180 kg/ha and fertilise heavily. With irrigation use 18–18–0 or 20–11–0 at the rate of 370–500 kg/ha depending on the phosphate level.

Make sure seed is covered, if necessary sow deeply to achieve this, and harrow or roll afterwards.

Winter carrying capacity and economic returns from nitrogen treated oats are largely dependent on good grazing management.

Oats should be about 25–30 cm high at the time grazing is commenced. When oats reaches this stage it should be grazed off quickly even though the feed might not be required at the time by the stock. This first autumn grazing should not be too severe and about 7–10 cm of growth should remain after grazing to ensure rapid growth.

More nitrogen may be an advantage after the second or third grazing.

Barley

Barley can be used for cutting or grazing in the same way as oats, but is a little more exacting in its requirements. It does not make such good use of second class soil as oats, nor is it so drought tolerant or resistant to waterlogging.

Sow in February or March at 70–100 kg seed per hectare into a well-prepared seedbed.

Wheat

Like barley, wheat is not as popular as oats, but it can be used in the same way. Wheat does well on heavy alluvial soils, where it can be grown in preference to oats for green feed and hay.

Rye and triticale

Rye and triticale may be used in the same way as oats for grazing. They are particularly hardy crops, and do better on light soils and in cold situations than oats.

Rye does not make good hay, and it should be grazed well before it heads since it becomes fibrous towards maturity. Cultural needs are as for oats.

Vetches

The use of vetches as a late winter-spring forage for dairy cows can be used to advantage. They have the ability to fill a major gap in the feed supply.

Vetches may be established successfully by sod seeding into a grass sward which has been kept short by heavy grazing or mowing. Light renovation before seeding may prove beneficial on some soils. The crop may be sown on a clean seedbed either alone or with oats. It will grow successfully on a wide range of soils, from light sandy soils to heavy clays. Areas of poor drainage should be avoided because water-logging will reduce the yield considerably.

Sowing may take place from March to late April, adequate soil moisture should be present when sowing.

Seeding rate is usually 25–30 kg/ha when sod seeding and increases to 50 kg/ha if broadcast.

It is most important to inoculate the seed with the correct type of inoculum. Neutralised fertilisers are essential when inoculated seed is sown in close contact with fertiliser — the fertiliser used should provide at least the equivalent of 250 kg/ha of superphosphate.

Up to 25 tonnes/ha green material can be obtained with a high protein content of from 16–20 per cent. The maximum feeding value is obtained about five months after sowing.

Fodder conservation

Feed requirements of cattle and sheep

Cattle and sheep need energy, protein, minerals and vitamins for maintenance, growth and reproduction, and for meat, milk and wool production. Under Australian grazing conditions, energy is generally the primary factor limiting animal production, with protein a close second in importance.

Digestibility is a measure of the percentage of feed actually used by the animal. It gives an indication of the energy value of a feed; hay and silage with a high digestibility has a high energy value. In late summer and autumn in southern Australia animals are unable to consume enough energy from the dry pastures for maintenance and production. In winter, when pasture is shorter than 15 cm, cattle are unable to satisfy their energy needs because they cannot eat enough pasture. When grazing lush, young pastures, cattle may also need more fibre.

Crude protein levels of above 12 per cent are essential for growth, production and reproduction in beef and dairy cattle and sheep. This can be supplied in good quality hay or silage.

Selecting a hay paddock

Good pastures make good quality hay. Hay paddocks should be selected in the autumn and prepared for hay cutting.

Weeds should be controlled and the pasture oversown with additional pasture species, particularly vigorous annuals, if necessary.

Sticks and stones should be picked up and removed from the area to prevent problems when mowing and the pasture should be topdressed to promote vigorous growth.

The paddock should then be grazed throughout winter as growth and weather/soil conditions allow, prior to the stock being removed and the paddock shut up at the beginning of the spring.

If the paddock has been repeatedly cut for hay in previous years it should also be topdressed with potash at this stage.

Hay and silage

Good quality conserved fodders can be made only from good quality pastures or crops. A good quality fodder is one that has a high digestibility of at least 65–70 per cent, a metabolisable energy level of 8.5 MJ/kg dry matter or higher, is readily eaten by sheep and cattle, and has a crude protein content above 12 per cent. A digestibility of 65–70 per cent is necessary because lower levels reduce the amount of hay or silage eaten, and reduced intake reduces production. The digestibility of well made hay and silage is similar to that of the original pasture or cereal crop.

In haymaking, the plants are preserved by drying to a moisture level at which bacteria and fungi are unable to grow. In silage making, plants are preserved at a higher moisture content in the acids produced during fermentation.

Compared with silage, hay has the advantage of being a saleable item, particularly if it is conserved as small rectangular bales, and it is also easier to feed out in wet weather than silage. As it is eaten in larger quantities than direct-cut silage made from the same plant material hay usually results in better animal production. However, provided the pasture is wilted or treated with formaldehyde (formalin) before it is ensiled, total yields of milk fat and protein from dairy cows are similar whether they are fed hay or silage.

Silage has the advantage that it can be made in damp weather and it normally costs less to produce and store per food unit than baled hay and total labour requirements for silage making should be less than for making and storing conventional baled hay. More than 50 per cent of the labour in conventional haymaking systems is involved in handling. The costs of storing and feeding silage exceed the costs for hay if tower silos are used, or if the silage is handled manually, but the use of silage grabs reduces silage handling costs considerably.

The objectives of successful hay and silage making are to:
- preserve the herbage with the minimum loss of nutrients
- produce hay or silage with a high feeding value
- incur the lowest possible cost.

Changes in digestibility of pastures

Stage of maturity

When several species are growing in a mixed pasture, cutting should take place when the dominant pasture species reaches the flowering stage, because after flowering there is rapid loss of digestibility in most pasture species (Figure 8.3).

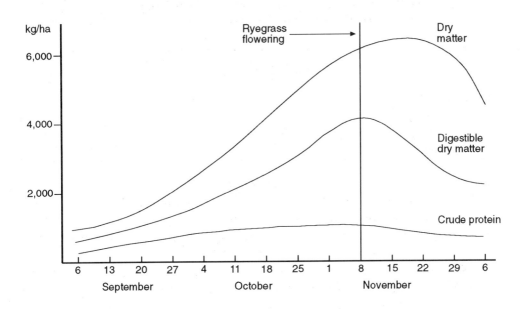

Figure 8.3 Yield of dry matter, digestible dry matter and crude protein of prennial ryegrass-sub clover pasture grown near Oakbank in the Adelaide hills

Table 8.1 shows that the ryegrass yielded the highest digestible dry matter when it was flowering. The highest dry matter yield — that is, the actual dry weight of the pasture per hectare — occurred two weeks after the ryegrass flowered.

Although slightly less pasture may be harvested at flowering, the quality of the hay or silage is much higher.

Table 8.1 Digestibility percentages of ten pasture species cut during spring in the Adelaide Plains

	Digestibility (%)									
Date	Barley grass	Annual ryegrass	Seaton Park sub clover	Barrel medic	Brome grass	Silver grass	Phal-aris	Lucerne	Currie cocks-foot	Demeter fescue
17 Sep	64*	72	71	72	70	69	67	72	60	67
1 Oct	60	76*	68*	65	72	66	64	65	56	53
15 Oct	53	59	62	57*	62*	61	59	62	57	59
29 Oct	46	52	54**	54**	56	59*	57*	62	56	60
12 Nov	42	48	49	47	47	54	52	58*	53	57
26 Nov	34	42	36	36	42	46	45	54	50*	55*
10 Dec	33	34	35	32	30	45	40	53	47	55

* Start of flowering   ** Seedpods descending

Table 8.2 Digestible dry matter yield, digestibility percentage and crude protein percentage for irrigated lucerne harvested at three, four, five and six week intervals at Milang

	Cutting frequency			
	3 weeks	4 weeks	5 weeks	6 weeks
Digestible dry matter				
Yield (kg/ha)	3700	5250	9750	11000
Digestibility %	67	66	72	67
Crude protein %	27	24	23	19

Species

Pasture species have different maturity dates. Annuals such as Seaton Park sub clover and annual ryegrass flower early in spring and should be cut at the flowering stage to obtain a fodder with a digestibility of about 70 per cent.

Table 8.1 shows just how much digestibility varies as plants mature.

Hay crops intended for sale usually have a perennial such as lucerne as the main component. Some perennials such as demeter fescue and lucerne do not decrease to such low levels of digestibility as do annual ryegrass, phalaris and sub clover late in the season when grown under dryland conditions. Lucerne is the main legume used specifically for haymaking. When grown under irrigation, its yield, digestibility and crude protein levels are high. Red clover is being used increasingly more for hay production as it is comparable to lucerne in some features.

Table 8.2 shows that a five or six week interval between cutting, which coincides with flowering, provides a high digestibility as well as the highest yield and an adequate level of crude protein.

Digestibilities of cereal varieties

The most suitable cereal varieties for southern Australian conditions are updated by the various state Departments of Agriculture. The wheat and barley varieties are selected for grain yield, while oat varieties are recommended for hay or silage production as well as grain yield and winter grazing. The recommendations vary for early, mid-season and late districts.

Hay or silage made from wheat, oats or barley is generally lower in digestibility than most pasture hays. Table 8.3 shows the digestibilities of some varieties of wheat, oats and barley.

It can be seen from this table that wheat, oats and barley plants decline rapidly in digestibility after the boot stage, but that wheat and barley may rise again in digestibility after flowering. The increase is caused by the formation of grain in the head which then has a high digestibility approaching 70 per cent. The increase offsets the rapid decline of digestibility in the stem and leaves. However, if cereal hay is cut too late, when the quality of the leaves and stems is very low, cattle eat only the heads and waste a great amount of straw.

Table 8.3 Digestibility percentages of wheat, oat and barley — varieties cut during spring and early summer on the Adelaide plains

Growth stages	Digestibility %									
	Wheat			Oats					Barley	
	Sword	Raven	Halberd	Glaive	Swan	Avon	Coolabah	Jaia	Ketch	Clipper
Boot	61	64	66	66	72	69	71	66	58	65
Flowering	56	55	56	51	55	60	69	60	55	59
Milk	57	57	59	54	56	58	62	59	56	57
Dough	58	58	62	60	53	53	61	58	56	60

Cut wheat and barley varieties at the milk stage and oats at flowering to take advantage of the highest digestibility. Table 8.3 also shows that some wheat and barley varieties still have digestibilities of about 60 per cent at the dough stage. Nevertheless, it is suggested that wheat and barley be cut for hay before the dough stage because the later cut hay may become infested with the hay itch mite on storage. The mite severely irritates the skin of anyone handling the hay and also irritates stock.

Changes in crude protein in pasture species and cereal varieties

Legumes such as clovers, medics and lucerne are higher in crude protein than grasses. Crude protein, like digestibility, also decreases as the plant matures. Early cut pasture hay has a higher level of crude protein than late cut pasture hay.

Crude protein levels in wheat, oat and barley crops at the time of hay and silage making are generally lower than in pasture species, especially at the flowering stage and later. Table 8.4 shows that barley varieties such as Ketch and Clipper are higher in crude protein than wheat or oats. In general, the crude protein levels in cereal hays are not high enough for optimum animal production (Table 8.5).

Table 8.4 Crude protein percentages of wheat, oat and barley varieties cut during spring and early summer on the Adelaide plains

Growth stages	Crude protein %					
	Wheat		Oats		Barley	
	Raven	Halberd	Swan	Avon	Ketch	Clipper
Boot	11.3	11.5	9.8	10.6	10.7	11.9
Flowering	6.8	8.3	4.6	6.9	12.4	9.9
Milk	5.6	6.6	6.0	6.1	10.1	9.6
Dough	7.9	7.1	6.0	6.2	8.9	8.3

Table 8.5 Changes of crude protein percentages in ten pasture species cut during spring on the Adelaide plains

Date	Crude protein %							
	Barley grass	Annual ryegrass	Trikkala sub clover	Barrel medic	Silver grass	Phalaris	Lucerne	Currie cocksfoot
17 Sept	13.3	9.7	18.7	23.3	14.9	13.0	24.1	13.6
1 Oct	11.1	7.8	16.1	20.1	12.6	13.1	21.4	11.7
15 Oct	10.0	7.3	14.8	21.8	12.4	13.6	22.5	11.6
29 Oct	8.4	6.8	12.3	20.1	10.2	11.9	21.2	11.2
12 Nov	6.4	4.9	9.2	14.5	8.2	8.1	16.6	9.2
26 Nov	6.9	4.3	8.5	13.2	6.4	7.1	15.7	8.5
10 Dec	4.7	3.6	6.9	9.4	5.3	5.1	14.1	6.9

The value of early cut hay and the risks of cutting hay early

It has been demonstrated that dairy cows fed early cut pasture hay with a digestibility of 70 per cent produced more milk and milk fat than cows fed late cut pasture hay with a digestibility of 60 per cent (Table 8.6).

Table 8.6 Milk and fat yields from cows fed early and late cut pasture hay at Northfield Research Centre

	Early cut hay 70% digestibility	Late cut hay 60% digestibility	Increase
Milk yield (litres/cow/day)	14.2	12.6	13%
Fat yield (kg/cow/day)	0.49	0.42	16%

High quality hay (with high digestibility and high crude protein) is produced from hay crops cut early in spring. However, when compared with late cut hay, hay cut early in the season has a greater chance of becoming rain soaked and mouldy. Studies have shown that it is still better to cut cereal hay early at the flowering stage and risk some rain damage than to cut later in the season. Mouldy cereal hay is marginally better than late cut hay.

Table 8.7 Yield and composition of milk from cows fed early and late cut cereal hay and mouldy early cut cereal hay at Northfield Research Centre

	Hay cut early	Mouldy hay cut early	Hay cut late
Production —			
Milk (litres/cow/day)	9.6	8.6	8.4
Milk Composition —			
Fat (%)	3.9	3.9	3.8
Protein (%)	3.3	3.3	3.3
Solids-not-fat (%)	8.7	8.5	8.7

Cutting guide — pastures

If the pasture contains a high percentage of clover it should be cut when the clover is still growing and still leafy. This stage should be before the grasses have set seed.

A mixed pasture should be cut when the main grasses come into ear and start flowering. Later cutting should be avoided, especially with annual pastures, as this will remove seed and so causes deterioration of the pastures although subsequent management can reduce this effect. Late cutting also reduces the chances of useful regrowth.

A greater bulk of hay can be obtained if it is cut late but the protein per cent in the hay is declining steadily and after about mid-flowering, depending on species, total protein yield is less.

Lucerne

Should be cut when it is about one-tenth in flower or when the basal shoots of the next growth appear. At this stage of flowering the plant is still leafy, the yield of protein per hectare is at its highest, and the stems are just beginning to form fibre. If the crop is cut near full bloom, half the weight of the plant is stems and fibre, the result is lowered protein content and digestibility; in fact, an inferior product.

Cereals

For the most palatable hay with the maximum digestible nutrients and yield, cereal hay should be cut at the early milky stage. However, oats may be cut up to the early dough stage.

The reduction in crude protein content of oaten hay from flowering to mature grain has been shown to be from 6.7–4.0 per cent on dry weight basis. Yield of dry matter and protein are at their maximum at the milky stage. Good hay should be free from weeds, stain and weather damage. If carefully cured the hay will show good colour, softness and a smal amount of dry, partly developed grain.

Sorghums, millets

Should be cut in the early to mid-bloom stage when protein content is at its peak. Mouldy sorghum fodder should not be fed to livestock as it is dangerous, especially to horses.

These species can produce problems in making hay. Sorghums particularly, because of their height and bulk, are often more suited to silage conservation.

Pasture management

Summary

For best quality hay observe these cutting times:

Pastures/clover dominant	— When clover is still leafy (about mid-flowering)
Pastures/mixed	— Early flowering of main grass species
Lucerne	— 10% of the plants in flower
Cereals Oats	— Flowering to early milky stage
Wheat	— Milky stage
Barley	— Milky stage with the beards still soft
Sorghum, Millets	— Early to mid-bloom.

Early cutting can produce an attractive, digestible high protein product and with careful curing and storing the usual spring surplus can be used to good advantage during the leaner months.

Aim to cut at the right stage of growth to raise the quality of the hay. When the right time is reached do not delay.

Even if one takes the risk of getting wet hay by cutting early, and is caught by rain, the resulting production should be at least as good as the best which can be achieved by cutting late.

Conserving fodder as hay

In haymaking, the crop or pasture with a dry matter content of 20 to 25 per cent is cut and then dried to 75 to 80 per cent dry matter to minimise deterioration. It is then necessary to package into a form convenient for handling and storage.

The haymaking operations — mowing

When a crop is cut for hay the moisture supply to the stem and leaves is cut off but, as the cut plants dry, nutrients are still being used by the cells in the plant. The cut plant material should be dried quickly to minimise the loss of valuable nutrients. The height of the cut above ground is important. A stubble length of 5–8 cm allows adequate air movement through the cut material and prevents moisture from being taken up from the ground.

Reciprocating mowers have been mainly used in the past. Slashers and horizontal rotary mowers of the multiple disc and drum types have become the more popular forage cutters. Reciprocating mowers block easily in thick pastures and leave ragged stubbles when driven at high speeds. They also cost more to maintain than do rotary mowers and if used on rough or stony land they wear quickly and break down frequently. Rotary mowers, although usually more reliable, need from three to six times the power of reciprocating mowers.

Potassium carbonate can be sprayed into lucerne at the time of cutting to improve drying rates. The chemical penetrates the waxy outer covering of the stem causing quicker moisture loss during drying. Lucerne can be ready for baling from 24 to 36 hours after cutting when weather conditions are suitable. A spray boom is mounted on the front of the mower and sprays the lucerne with a two per cent potassium carbonate solution at the rate of approximately 250 litres of solution a hectare, just before the herbage is cut. The tank for the chemical is mounted on the front of the tractor or trailed behind the mower. A diaphragm pump driven by a hydraulic motor may be used to deliver the chemical to the spray boom.

Conditioning

Conditioners crimp or crush the stems of the plant and increase the rate of drying of hay crops by 60 to 80 per cent. By comparison, the best rotary mowers improve drying rates by only ten per cent compared with a reciprocating mower.

Crimpers or conditioners tend to be least effective in their crushing and bruising operation in heavy crops. In such crops the windrows are dense and air movement through them is poor. Conditioners do not perform well in uneven crops. If the soil is damp and the stubble long, conditioners may pull crop residues from the ground, and soil may build up on the rollers. Although conditioners reduce drying time, they also increase the risk of losing nutrients through leaching if rain falls after cutting.

Mower-conditioners mow and condition in the one operation and are ideal for lucerne crops. The thick lucerne stems are crimped and dry at a similar rate to the leaves. This may allow earlier baling, provided the material is not packed into tight windrows where drying may be slowed by poor air circulation.

Raking

The mown plant material should be raked into windrows after initial drying of the crop, but before the leaves become brittle. Nutrient losses due to rain are less from plants in windrows than from a mown unraked crop, but windrows are harder to dry out after heavy rain. If rain is threatening it is best to keep the windrows small. Windrows may have to be turned onto dry ground with a rake if they do not dry underneath or if they become wet because of rain.

Finger wheel and bar reel rakes are the two most widely used types but the type of rake is not as important as the time and manner of raking. Raking evenly and in one direction for quick and even drying and ease of baling later is most important. Raking too early after mowing should be avoided as a tight ropy windrow will be produced and on hot days, raking early in the morning or in the evening, when there is dew on the windrows, will reduce leaf loss. Travelling too fast will also cause unnecessary leaf loss.

Baling rectangular bales (20-25 kg)

Hay should be baled when it reaches a dry matter content of 75 to 80 per cent. When the dry matter content is below 75 per cent moulds may grow and reduce the nutritive value of the hay, and excessive heat, which results in a serious risk of spontaneous combustion, may develop in the stack if the hay is carted too soon.

Two simple tests can be used as a guide to the right time to bale. For the first test, pull a few stems of plant material between a tightly pressed finger and thumbnail. The hay is ready to bale if moisture does not appear on either the finger or thumbnail. In the second test examine the nodes of the grass species in the cut material and if they are withered and dry the material is ready to bale. Moisture meters are now a common piece of equimpent used to test hay for correct moisture levels.

Handling and storage of small rectangular bales

Wet bales dry best when they are stooked in pairs. When bales dry quickly nutrient losses and mould growth are minimised so the extra labour that stooking requires is worthwhile. Stooked bales will also shed any further rain.

Bales are traditionally handled with pick-up bale loaders and elevators into the stack. Manual stacking systems require about 2.5 manhours per tonne of hay handled from paddock to stack to feeding out.

Hay to be stacked should have a dry matter content of at least 75 per cent and preferably 80 per cent; small wet pockets in the stack should be avoided to prevent spontaneous combustion.

The following points should be adhered to for successful hay stacking:
- Ensure that the site for the stack is level, well-drained and safe from fire. Build the stack end-on to the prevailing wet weather.
- Give outside bales in the bottom layer a slight tilt upwards by packing loose hay under the outer edge.
- Use only good-sized and well-shaped bales for the corners.
- Keep the sides perpendicular and avoid protruding or receding bales.
- After the bottom layer has been completed, complete the ends of each succeeding layer before the centre. The last bales to go in each layer must be those in the centre. Take care not to jam large bales in the centre as this may dislodge the edges.
- Stack to a pattern. There are various patterns for building different layers. Figure 8.3 shows a plan that has proved satisfactory.

Experiments have shown little effect of storage on the digestibility of pasture, oaten and lucerne hays when kept under cover for three years. Small losses of crude protein were observed only in the pasture hay.

A number of models of automatic bale stacking wagons are available. These wagons take up to about 100 bales. The load can be automatically

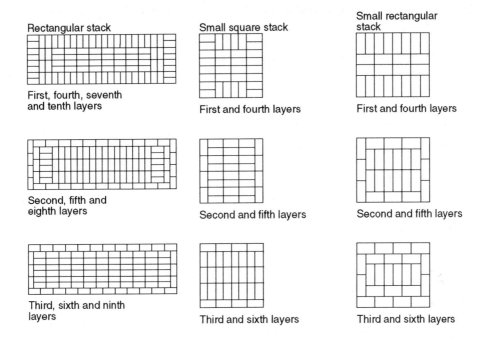

Figure 8.3 Stacking patterns for various layers of different shaped haystacks

stacked by tilting the wagon either in a hayshed, which has a back on it, or in the paddock where the stability of the stacks depends on the ability of the operator to support the back of the first load suitably by the use of a pyramid of bales. The wagon can also be used to automatically unload individual bales into a hayshed. This type of equipment is used almost exclusively by specialist lucerne growers.

Front-end loaders are now being used to handle groups of eight or ten bales at a time. Groups of bales in the flat 8 configuration (four rows of two) are arranged in groups with either a platform behind the baler or on a sled. In each type of machine, once a group of eight bales is formed it is automatically released onto the ground. If bales are damp they do not slide into place easily.

A grab on the front-end loader, with prongs that pierce the eight bales, is used to pick up the bales and load them onto a trailer. One man can load 240 bales onto a trailer in 50 minutes, although some rearrangement of the bales in the stack may be necessary. The hayshed should preferably have sides to support the bales.

A flat 10 system has been developed for handling small rectangular bales. The bales are arranged in a group of ten on a platform behind the baler. The first two bales are moved to the back of the platform and the remaining eight bales are positioned into four groups of two at right angles to the first bales. The bales are automatically released onto the ground and are picked up with a grab similar to the one used for flat 8s. The arrangement of the group of ten bales has the advantage that layers of bales can be tied when stacked mechanically on a truck.

Large hay packages

The high input of labour needed to handle small rectangular bales and the cost and scarcity of labour during haymaking has led to the development of machinery capable of producing large hay packages. The large packages require less labour to produce and all the hay handling is fully mechanised

Claims are made that these large hay packages are weather-resistant and that only small losses of nutrients occur from the bales when stored in the open. However, work in South Australia has shown that half the nutrients are lost from 100 kg fodder rolls when stored in the open for over nine months. In the USA it has been found that large compressed haystacks and large round bales stored in the open lost up to 12 per cent of their nutrients.

Fodder rollers producing bales weighing about 100 kg have been the most popular type of large balers used in South Australia in the past decade. This method of baling may reduce costs by 30 to 50 per cent when compared with conventional haymaking, but when stored in the open, these rolls do not weather well. Experiments at Bordertown showed that fodder rolls lose over 50 per cent of their digestible dry matter when stored in the open for nine months, and the soil type on which the rolls are stored has no effect on the rate of loss. Fodder rolls are unsuitable as a drought reserve and should be fed out in the summer and autumn immediately after haymaking.

Machines which produce round bales weighing 500 to 750 kg at the rate of 7 to 9 tonnes per hour are now very popular. These bales have a density of 120 to 160 kg/m^3. These balers roll the hay between belts or slats in the machine rather than on the ground like the small fodder roller. These large rolls should not be stored in the open touching one another as moisture will be trapped between them and will lead to deterioration of the hay. Losses of up to eight per cent have been recorded after six months storage in the open. The rolls can also be stored in the shed stacked end on end up to three high. They are handled with a hydraulic grab on either the front or the back of a tractor.

Although not very common any more, labour savings of up to 75 per cent can be made when stacking wagons are used instead of conventional balers. Machines available produce compressed stacks with a weight of one to six tonnes. The five tonne machines have a capacity to make 12.5 tonnes of hay per hour, while the one tonne machines will make six tonnes per hour. An extra stack moving machine is required for some models of stackers. The labour required, and hence the cost of labour, decreases proportionately as the size of the stacking wagon increases.

The field equipment needed to produce windrows for the large stacking wagons is the same as that used for producing windrows for conventional balers. The moisture content of hay in the windrows for stacking should be slightly higher than that for rectangular bales. If the windrows are too dry, the stacks produced will be light and unstable and therefore susceptible to wind damage. Stacks of cereal hay are prone to wind damage because they lack cohesion. Stacking in the morning or evening, and storing near windbreaks such as trees will minimise losses.

Stacks should be left to settle for at least two days before they are moved because they tend to fall apart if they are lifted and moved immediately after stacking. Losses of 12 per cent of the digestible dry matter of hay from compressed stacks over five months have been recorded.

Balers which produce large rectangular bales weighing 450 kg and have a baling capacity of seven tonnes per hour are now being marketed. They produce bales of an average density of 90 kg/m^3. The baler picks up the windrow and rolls small bales which are pressed into a cage to form one large rectangular bale tied with three strings.

These bales, if left in the open, appear to be more susceptible to damage due to weather than compressed haystacks or large round bales. However, their rectangular shape allows them to be stacked three high and they can be put into conventional hay sheds where damage due to weather is reduced.

Feeding hay

Very late cut pasture hay with a low digestibility is not readily eaten by cattle and often much of it is wasted. The opposite is true of pasture hay cut early in the season. In general cattle waste a greater proportion of cereal hay than pasture hay because they select the heads of late cut cereal hay and leave the stems which are of low digestibility.

Small rectangular bales are best fed in hay racks so that wastage due to trampling and fouling is reduced. Large rectangular bales should also be fed in a rack because animals spread the hay if it is paddock-fed.

Whether making or buying hay Table 8.8 on the next page provides a clear guide to quality. Always aim for the best.

Cattle will pull hay from the stacks and trample it into the ground. Losses of 35 to 45 per cent of the dry matter have been observed. The use of feeding panels or an electric fence can reduce the feed wasted to 5 per cent.

A big proportion of large round bales is also lost if they are spread out and fed in the paddock. The use of feeding panels reduces these losses also.

Rules to remember

❀ Cut pasture for hay at the early flowering stage.
❀ Do not bale hay that is too wet.

Table 8.8 Grades for hay quality

Quality	Description of Hay
Excellent	Early cut. Good green colour, leafy. Well cured with little or no weather damage. Leaf entire. Soft feel, sweet (free from mustiness). For meadow hay — a high clover content.
Good	Early cut hay of all kinds, cured with little or no weather damage or sweating. Leafy and green. Leaf in good condition. For meadow hay — a fair clover content.
Fair	Early cut hay of all kinds that suffered more than slight weather damage or sweating. Also later cut hay but not overmatured, well made, with little or no weather damage. Both tend to be harsh or stemmy, with reduced green colour. Often musty. May have some mould. For meadow hay — grassy, low clover content.
Poor	Early hay, severely damaged. Later hay, weather damaged or sweated. Overmature, stemmy and seedy hays. Little or no green colour. For meadow hay — little or no clover.
Bad	Hay of all kinds, cut at any stage of maturity, but so severely damaged by weather and sweating as to be musty, mouldy, badly stained, severely bleached or discoloured. Extremely overmature hay.

Irrespective of the quality classifications, individual samples may be altered to other grades for such things as weeds, or other foreign material, dustiness, insect, vermin or other damage.

- Cart hay soon after baling.
- Stack hay under cover if it is to be stored more than a few months.
- Feed hay in racks, behind panels or use electric fences.
- Devise a system to minimise the amount of labour required for the total operation.

Conserving fodder as silage

The aim of storing green crops as silage is to preserve the material with minimal loss of nutrients. The first important step is to achieve and maintain oxygen-free conditions which prevent the loss of nutrients and the wrong type of fermentation. The second step is to prevent protein destruction by bacteria called clostridia which can grow in an oxygen-free environment.

Clostridial activity in silage can be inhibited by raising the dry matter content of the herbage to 30 per cent by wilting before ensiling, or by allowing a natural acidic fermentation to occur. Bacteria normally present on the herbage or applied through inoculation convert plant sugars to lactic acid and acetic acid, both of which inhibit the growth of clostridial bacteria.

The presence of oxygen will allow the breakdown of plant sugars into carbon dioxide and water and the outward sign of this is a rise in temperature of the ensiled crop. The more rapid and greater the rise in temperature, the greater is the loss of nutrients. If the temperature rises above 45°C, the digestibility of the protein is reduced.

Therefore, to reduce the loss of nutrients in the silage stack, it is necessary to rapidly fill the stack, effectively compact the herbage, and seal the stack immediately.

Cutting herbage for silage

The first step in silage making is cutting and lacerating the herbage to be ensiled. Factors which influence the choice of suitable machinery for harvesting forage on a particular farm include the quantity of silage to be made, the amount of money available, the distance between field and bunker, the number and types of tractors available, the type of crop to be ensiled and the type of land on which the crop is grown.

There are three basic methods of cutting and carting the crop. The advantages and disadvantages for each are set out in Table 8.9.

The purchase of new equipment to make less than 300 tonnes of silage cannot be justified. A buck-rake, if on hand, may be suitable for making small quantities of silage if the distance to the stack is short. A forage harvester and large trailer should be used for long hauls.

Stemmy crops such as maize and sorghum must be chopped with a double or precision chop forage harvester, so that there is adequate compaction in the silage stack.

Chopping and lacerating the crop allows ready access of bacteria to the plant sugars, with the result that fermentation is rapid and the silage is of high quality. This can be achieved only with a forage harvester. The three types of forage harvesters available are: the flail machine, the double-chop machine and the precision chopper.

A flail machine is simple, robust and cheap, and produces long pieces of lacerated grass which are more difficult to remove from the silage stack than grass cut with a double-chop machine. The flail machine (without a trailer) requires a tractor of 19–26 kW for a one with a 1 m cutting width.

The double-chop machine also cuts with flails but the cut herbage is further chopped with a flywheel or cylindrical chopper. The degree of chopping can be adjusted. The chopped herbage compacts readily so more can fit into the trailer and the amount of rolling needed to compact the herbage in the silage stack is reduced. The machine is expensive to buy and has a higher power requirement than the flail machine.

The precision-chop machine has a flywheel or cylindrical chopper to chop the mown crop. Some models of these machines also have a sickle-bar mower

Pasture management

Table 8.9 Advantages and disadvantages of the three commonly used silage making techniques

Equipment	Advantages	Disadvantages
Mower and buck-rake.	Low capital outlay.	Slow rate of work and danger of deterioration of mown crop and herbage in silage stack.
	Low maintenance and running costs.	Two operations and inefficient use of labour.
	Low powered tractor is suitable.	Suitable to make only small amounts of silage.
	Allows wilting.	Ground must be smooth.
		No chopping or laceration.
		Distance from crop to stack must be short.
		Self-feeding is difficult.
Forage harvester.	Efficient use of labour.	Does not allow wilting.
	One operation only.	High capital outlay.
	Mechanical application of additive is possible.	High powered tractor is required.
	Chops the herbage.	
	More suitable for rough ground.	
Mower and forage harvester.	Efficient use of labour.	Two operations.
	Allows wilting.	High capital outlay.
	Mechanical application of additive is possible.	High powered tractor is required.
	Chops the herbage.	Requires smooth ground.
		Danger of crop deterioration.

attachment. These machines are very expensive to buy and maintain, but have a high working rate. With special attachments they are particularly suitable for such crops as maize and sorghum.

All three machines are suitable for picking up cut, wilted herbage. Forage harvesters differ in the width of cut (from 1–2 m), the type of attachment to the tractor (three-point linkage or trailed), the line of attachment (in-line of off-set machines), the number and type of flails and their mode of attachment, the designated rotor speed (1400–2000 rpm), and the height of discharge (from 1.4–3.0 m).

The output of a forage harvester depends on the tractor power and the system of operation. The figures in Table 8.10 are an approximate guide only and vary according to the weight of crop per hectare.

Cutting as close to the ground as possible usually results in contamination of the silage with soil, and slow recovery of the crop. A stubble of approximately five centimetres should be left after cutting.

Table 8.10 Power requirements and output for various sizes and types of forage harvester

Type of forage harvester	Tractor power (kW)	Output (hectares/hr)
1 m flail	under 30	0.24–0.32
1 m flail	over 30	0.36–0.49
1.2–1.5 m flail	over 30	0.49–0.65
1.5 m double-chop	over 30	0.41–0.73
Precision-chop	over 30	0.57–0.97

Wilting

The higher the dry matter content of the crop cut for silage, the more the ensiling fermentation is inhibited and the more compaction is required. Silage in which a small amount of fermentation has occurred is consumed in greater quantities by sheep and cattle and thus has a higher nutritive value than silage that has undergone extensive fermentation.

A wet, lush crop loses nutrients in the moisture draining from the stack and encourages fermentation by clostridia bacteria. Such drainage can also pollute water courses if the effluent reaches them. Avoid excess moisture in the stack by wilting the crop, and by cutting it when herbage is free from rain or dew.

If the herbage is to be wilted it can be cut with a rotary or disc mower. A rotary mower has the advantage of windrowing a wide cut into a width suitable for the forage harvester.

Although wilting adds an extra operation to silage making, it is possible to complete mowing and gathering in the same time as it would take to cut directly with the forage harvester. This is because the effective cutting width of the mower can be greater than the width of the forage harvester and yet still form a windrow small enough to be picked up by the forage harvester. In addition, when picking-up windrows, forage harvesters can travel at approximately twice the speed possible when cutting a standing crop.

A further advantage of wilting is that it greatly increases the weight of dry matter that can be carried in the trailer.

The herbage should be wilted to a dry matter content of 30 per cent. A lush pasture should be left to wilt for 12 hours under hot, dry weather conditions and up to 48 hours under cool, overcast conditions.

The dry matter content of the pasture may be estimated by the "grab test". Chopped herbage is tightly squeezed into a ball between the hands for about one minute. If the hands have juice on them when unclasped, the herbage has a dry matter content of less than 25 per cent. If the ball of herbage holds its shape and the hands are not wet, the dry matter content is between 25 and 30 per cent. If the ball falls apart, especially if it springs apart; the dry matter content is above 30 per cent.

Wilting herbage to a dry matter content above 30 per cent will make compaction in the stack difficult, thus encouraging mould growth and lowering digestibility.

Silage additives

Chemicals may be added to herbage to help preserve it during storage. Although a large number of additives have been recommended overseas, many are expensive and their use cannot be economically justified in Australia. However, some are comparatively cheap, readily available and may improve the preservation and feeding value of silage made from crops with a low sugar content such as lucerne.

The main additives available are formalin, molasses and formic acid.

Formalin (40 per cent formaldehyde solution) has been tested as a silage additive. Formaldehyde inhibits the activity of silage bacteria so there is less fermentation and sheep and cattle will eat more silage. Experiments have shown that cows fed formaldehyde silage produce 2.4 litres more milk per day than cows fed untreated silage. Formalin is applied at the rate of 2.3 litres per tonne of fresh herbage with a pump driven from a belt and pulley attached to a wheel trailing behind the forage harvester. The cost of formalin is approximately $1.00 per tonne of silage. Formalin should be handled with care as it irritates the eyes and respiratory system.

Molasses works by providing readily available sugar which acts as a food for the beneficial lactic acid bacteria. The lactic acid formed prevents the growth of spoilage bacteria. However, molasses is messy to handle, difficult to apply and expensive. If used, it should be applied at the rate of 14 litres per tonne of fresh pasture. The cost of molasses is about $3.00 per tonne of silage.

Formic acid, which restricts the growth of spoilage bacteria, is applied as an 85 per cent solution at the rate of 2.3 litres per tome of fresh herbage. However, like molasses, formic acid is expensive, costing approximately $3.00 per tonne of silage. The acid is very toxic and corrosive, and causes irritation to the eyes, nose and throat and severe burns if it comes in contact with the skin. Rubber gloves, goggles and protective clothing should be worn when the acid is handled.

Carting

For an efficient silage making system the trailer capacity should be large and the distance from the crop to the silage stack should be as short as possible. The weight of herbage that can be carried by the trailer depends upon its volume and the density of the load.

Two important features of a silage trailer are large capacity and a mechanism for rapid unloading. A V-shaped trailer will probably hold no more than 0.5 to 0.7 tonnes under normal working conditions with a flail forage harvester, however, a hinged gate at the back allows rapid unloading.

Box-type trailers are slower to unload because they involve chains and slats moving along the floor. However, they have a capacity of 4.5 to 12.5 m^3 and can carry from 2 to 5 tonnes. Some box trailers also have a side-delivery mechanism enabling the green crop to be fed out directly into troughs.

A suitable silage trailer may be constructed from an existing trailer or from a tip-truck. The tray and chassis of an old tip-truck may be used as the basis of a four wheeled trailer. It is important to incorporate a rapid unloading mechanism.

Having the bunker or pit close to the crop being cut reduces the costs of silage making. The costs of feeding silage will also be reduced if a bunker site suitable for self-feeding is chosen.

Storing silage

Methods of storage range from an inexpensive uncovered silage heap on top of the ground in which up to 70 per cent of silage may be wasted, to an expensive glass-lined Harvestore® tower silo where losses are low (see Table 8.11).

High losses can be expected with the **uncovered heap** method of storage, because it is difficult to compact the herbage and prevent the entry of oxygen.

The capital cost of **vacuum tents** is approximately $5 per tonne of silage and usually these tents cannot be used for more than one or two seasons. It may be impractical to pump all the air out of the silage stack as the tents can hole easily. Economically the vacuum tent cannot be justified.

The **tower silo** is usually constructed of galvanised steel and is glass-lined or treated with protective paint to make it airtight and acid-resistant. The capacity may be in excess of 300 tonnes. Chopped fodder is blown into the top of the silo and unloading is highly mechanised. Although losses of dry matter are low, the capital outlay for tower silos is difficult to justify in Australia, particularly where the silo may be used only once annually.

Another method of silage production is to produce it in large **round bales** similar to hay. The silage is cut and wilted for a short period. When the moisture content is at the correct level it is inoculated with bacteria essential for the ensilation process and baled in large round bales which are finally wrapped in plastic to exclude air. The advantages of these bales are that they can be handled easily, and usage of one does not spoil the rest of the silage. Silage made in this manner is expensive, but given that livestock perform better on silage than hay the additional costs are returned in the additional profit from the livestock. Silage made in large round bales require the normal equipment to make large round bales of hay as well as a wrapper for individually wrapping bales or a wrapper for wrapping bales together in a long tube. Individually wrapped bales need to be handled with forks that do not damage the plastic. Tube wrapped bales are handled before wrapping occurs. One of the major problems with this type of silage

Table 8.11 Advantages and disadvantages of various ensiling techniques

Type	Advantages	Disadvantages
Stack An above-ground heap which is consolidated, but without sides. It may or may not be covered.	No capital cost. Drainage problems minimised. Self-feeding possible.	Poor quality silage. Losses may be as high as 75 per cent. Unsafe to consolidate.
Vacuum tent An above-ground cylindrical heap covered with a plastic tent which is evacuated.	Low losses if well made. Drainage problems minimised.	Stacks less than 40 tonnes. High cost of plastic tents. Difficult to self-feed. High losses if plastic is torn.
Walled clamp and bunker An above-ground heap between two near-vertical walls. Either wedge or run-over. Covered or uncovered and consolidated by tractor.	Low losses if well made. Drainage problems minimised. Safer to consolidate. Large volume possible. Self-feeding possible.	Some capital cost. High waste if uncovered.
Round Bales Silage wrapped in plastic in bales similar to large hay bales.	Low losses. Easily stored. Only open small portions at a time. Uses haymaking machinery.	High capital cost in machinery. Plastic easily pierced.
Pit Below-ground heap which is tractor consolidated. Bare soil or lined near-vertical sides. Lined or bare bottom. Covered or uncovered.	Low losses. Safer to consolidate. Large volume possible. Self-feeding possible.	Some capital cost. Drainage problems, unless sited on side of hill.
Tower Silo Concrete or lined metal tower. Loaded with blower. Unloaded either by hand or mechanically. Mechanical unloaders may be at the top or bottom of the silo.	Low losses. No consolidation problems. Automatic unloading.	High capital cost. Unloading machinery may jam if silage is too wet. Maintenance is expensive.

production is that small holes in the plastic wrapping allow air to enter which will spoil the silage. Birds and mice are particularly a problem by putting holes in the plastic.

The high losses of dry matter which occur with uncovered silage stacks considerably increase the cost of producing each tonne of silage. The use of a plastic-covered **walled clamp and bunker** reduces silage losses and is an economical method of storage.

The dry matter losses of pit silage are low particularly if the pit is lined and covered. If a two-ended pit is cut through the crest of a hill then filling and mechanical feeding-out is easier.

Pits should be sited on ground which is well drained, preferably on the side or through the crest of a hill. Excavation costs can be depreciated over many years. The floor of the pit should be slightly mounded down the centre with an outlet fail of 1 in 50 along the crest of the mound. Drainage tiles may also be used along either side of the floor. The floor must be firm under wet conditions and therefore may be covered with concrete, gravel, or sleepers or may be lime-stabilised. The walls should have an outwards slope of 10 cm for each 80 cm of height. They should be smooth and may be lined with concrete. The width of the pit should be 4 metres or more. Proportionately less silage is wasted if the walls are high. The ends of pits in flat ground should have a gentle downward slope which can easily be negotiated with a light tractor; a slope of 1 m in 5 m is the maximum. The excavated soil can be heaped along the sides to increase the height of the sides.

A walled clamp should also have a firm mounded floor with a slope of 1 in 50. The walls are usually made of concrete or timber and should be smooth and at least two metres high and preferably higher. They need to be strong to withstand the enormous pressures of the silage and to ensure the safety of the tractor driver consolidating the clamp. The walls should slope outwards at the rate of 1 in 8.

A pit 5.5 metres wide at the bottom, 6.7 metres wide at the top, 2.4 metres deep and 6 metres long will hold approximately 100 tonnes of silage (see Table 8.12).

Recent research has shown that the quantity of air initially included within the ensiled crop is of little significance because the oxygen is quickly used by the plant material. It is more important to prevent further fresh air from entering. The main way in which air enters is by suction due to convection currents caused by heating in the heap. It is important to prevent this inflow of air by consolidating the silage and by quickly sealing the sides

Table 8.12 Tonnage capacities of silage pits and bunkers.
(Based on 1.6 cubic metres per tonne.)

Width at bottom	Width at top	Depth	Length of main section of pit or bunker (For total length add 10 m for 5 m slope at each end).			
			6 metres	12 metres	18 metres	24 metres
Metres	Metres	Metres	Tonnes	Tonnes	Tonnes	Tonnes
3.7	4.6	1.8	53	82	112	141
3.7	4.9	2.4	73	114	154	195
4.3	5.2	1.8	61	95	128	162
4.3	5.5	2.4	83	130	177	224
5.5	6.4	1.8	76	119	162	203
5.5	6.7	2.4	104	163	220	278

and top of the pit or clamp. One end of the pit or wedge is filled first. Each night, plastic sheeting should be drawn over the silage.

The cut herbage should be dumped directly onto the silage stack from the forage trailer by either driving it onto the stack or by backing the trailer to the side of the pit. Alternatively, the herbage can be tipped from the trailer near the stack and is then loaded onto the stack with a tractor and buck-rake.

If the Dorset Wedge system is used and the herbage is finely chopped it is not necessary to give much additional rolling to the silage heap with a tractor. Driving back and forth with the tractor and buck-rake continuously, while filling the stack, will consolidate the heap as much as is necessary. Dry, mature, or long herbage requires more consolidation than wet, immature or short herbage.

When driving a tractor on the heap remember the following rules for tractor safety:
- Use an experienced driver.
- Select low gear before driving onto the stack; use the clutch gently while on the stack.
- Do not pile the herbage above the top of walls.
- Have strong walls which can support the weight of tractors.
- Always back up a ramp when consolidating or carrying a buck-rake.
- Set the wheels as wide as possible.
- Consolidate the sides thoroughly so that the whole surface is firm.
- Use a tractor with a safety frame.

A thin polythene covering will more than pay for itself by reducing the amount of silage wasted. A 250 tonne clamp or pit can be covered at an approximate cost of 15 cents per tonne. The polythene sheeting should extend well down the sides of the bunker between the walls and the silage. The polythene should then be weighted securely with earth or old tyres.

It is important to use thin cheap, black polythene because sunlight makes it brittle within eight months and so the polythene can be used only once.

Feeding silage

Beef cattle can gain liveweight at the rate of 0.8 kg per day if fed a silage of high feeding value. However, for liveweight gains in excess of 1 kg per day and for maximum milk production in dairy cows, silage should be supplemented with a grain concentrate.

Sheep fed direct-cut silage may only maintain condition. If fed lucerne silage made from wilted herbage or herbage treated with formaldehyde or formic acid, liveweight gains up to 0.1 kg per day are possible.

Since 75 per cent of the weight of silage may be water, the weight of silage fed should be four times the weight of hay fed.

The efficiency of a silage system depends on how the silage is fed out. The use of hand forks and silage knives is laborious and inefficient.

Mechanical grabs should be used to remove silage from pits and clamps or the silage should be self-fed. Self-feeding results in less smell than where the silage is fed out in the paddock with mechanical grabs.

Front-end and three-point-linkage **mechanical grabs** operating from the hydraulic systems of a tractor are ideal for removing silage from the heap. Front-end grabs are more versatile and safer to use than rear-mounted grabs. Up to 250 kg of silage can be picked up in each grab. The silage is fed direct to the stock or loaded onto a flat-top or forage trailer and then fed out.

The ground around the silage heap should be firm so that the tractor does not get bogged.

Self-feeding reduces the labour and cost of feeding silage. It is particularly suitable for feeding silage to mature cattle. Self-fed silage for young cattle and sheep should be finely chopped, otherwise they may have difficulty in pulling it from the stack.

The width of silage face must be sufficient for the number of stock being fed. For mature cattle with 24-hour access, this width is 15 to 20 cm per head; with 8 to 12-hour access it is 25 to 40 cm per head; and for more restricted access 45 to 60 cm per head. For sheep, 3.5 to 4.5 m per 100 sheep is adequate when there is 24-hour access. An adjustable rail barrier or an electric fence should be placed along the feeding face to restrict the access of stock to the face. A solid base or platform should be provided at the face of the silage stack to give stock a firm footing and to reduce silage waste.

Silage making systems

Silage making should be carried out on a large scale if it is to be a successful fodder conservation practice.

The silage making system used and the combination of machinery purchased will require careful planning. The purchase of new machinery is probably not justified if less than 300 tonnes of silage is to be made annually. The amount of silage to be made, the labour available, whether the herbage will be wilted or cut directly and the degree of chopping, are factors to be considered when planning a silage making system.

The capacity of the silage making machinery should match the quantity of silage being made, otherwise the costs will be excessive. Considerations should take into account the amount of time available to make this silage. High output systems generally need specialised equipment.

The number of tractors available determines the number of simultaneous operations which can be performed.

Wilted pasture is likely to make better silage, but it must be ensiled quickly and sealed immediately. If the pasture is wilted, a mower and an extra operation are required, although extra time may not be involved.

The length of grass ensiled affects its consolidation and may limit the method of feeding, especially with self-feeding. Finely-chopped grass can be achieved only with a double-chop or precision-chop forage harvester.

List of species

Pasture species

Annual ryegrass	*Lolium rigidum*	Puccinellia	*Puccinellia ciliata*
Balansa clover	*Trifolium balansae*	Red clover	*Trifolium pratense*
Cocksfoot	*Dactylis glomerata*	Rose clover	*Trifolium hirtum*
Italian ryegrass	*Lolium multiflorum*	Sarradella	*Ornithopus compressus*
Kangaroo grass	*Themeda australis*		
Kikuyu	*Pennisetum clandestinum*	Strawberry clover	*Trifolium fragiferum*
		Subterranean clover	*Trifolium* spp
Lucerne	*Medicago sativa*	Tall fescue	*Festuca arundinacea*
Medics	*Medicago* spp	Tall wheatgrass	*Thinopyrum elongatum*
Paspalum	*Paspalum dilatatum*		
Perennial veldt	*Ehrharta calycina*	Vetch	*Vicia* spp
Perenniel ryegrass	*Lolium perenne*	Wallaby grass	*Danthonia* spp
Phalaris	*Phalaris aquatica*	Weeping grass	*Microleana stipoides*
Prairie grass	*Bromus catharticus*		

Pest species

Barley grub	*Persectania ewingii*	Pasture cackchafer (yellow)	*Sericesthis geminata* and *S. planiceps*
Blue-green aphid	*Acyrthosiphon kondoi*		
Cabbage centre grub	*Hellula hydralis*	Pea aphid	*Acyrthosiphon pisum*
Climbing cutworm	*Helicoverpa punctigera*		
		Pink cutworm	*Agrotis munda*
Field cricket	*Teleogryllus commodus*	Red-legged earth mite	*Halotydeus destructor*
Pasture cackchafer (black)	*Aphodius tasmaniae*	Sitona weevil	*Sitona discoideus*
		Spotted alfalfa aphid	*Therioaphis trifolii*
Pasture cackchafer (red)	*Adoryphorus couloni*		

Weed species

Barley grass	*Hordeum leporinum link*	Innocent weed	*Cenchrus pauciflorus*
		Noogoora burr	*Xanthium pungens*
Bathurst burr	*Xanthium spinosum*	Potato weed	*Heliotropium europaeum*
Broad-leafed dock	*Rumex obtusifolius*		
Caltrop	*Tribulus terrestris*	Salvation Jane	*Echium plantageneum*
Cape tulip (one-leaf)	*Homeria breyniana*		
Cape tulip (two-leaf)	*Homeria miniata*	Silver grass	*Vulpia* spp
Capeweed	*Arctotheca calendula*	Skeleton weed	*Chondrilla juncia*
		Soursob	*Oxalis pescaprae*
Clustered dock	*Rumex conglomeratus*	St John's wort	*Hypercium perforatum*
Curled dock	*Rumex crispus*	Storksbill	*Erodium* spp
Fiddle dock	*Rumex pulcher*	Swamp dock	*Rumex browii*
Horehound	*Marrubium vulgare*	Three cornered jack	*Emex australis*

Index

A
Acid soil, species 51, 61, 64
Acidity *see* pH
Aeration 74, 77–8, 137
Agro plow 13, 77
Alkaline soils, species 51, 52, 55, 64
Aluminium 61, 78
Annuals 10
Arability 11, 80

B
Bacterial wilt 67
Baker boot 81
Balansa clover 57, 60
Barley 88, 150, 156, 160, 161
Barley grass 88, 107, 110
Barley grub 88–90
Bathurst burr 113, 114
Biennials 10
Biological control, pests 85, 87–8, 97, 103, 104, 106
   weeds 117–19
Bloat 46, 69
Blue-green aphid 56, 67, 90–1, 99
Brassica crops 77, 150, 151–2
Bray test 142
Broadcasting 16, 34, 51, 82–4

C
Cabbage centre grub 91–2, 152
Calibrating 78–9, 129–30
Caltrop 112, 114
Cape tulip 107, 108, 111, 116, 126
Capeweed 107, 110, 116, 122
Cereal crops 9, 37, 48, 77, 116, 160–2
   pests 88, 100
Cereal root disease 10, 46
Certified Seed 17–18, 114
Chemical ploughing 79, 82
Chou moellier 77, 91, 92, 150, 151–2
Cleaning crops 75–6, 116–17
Climate 7-10, 12
Climbing cutworm 93–4, 100, 152
Clothing, protective 132–3, 174
Clover disease 55
Clover scorch 55, 57
Clovers 45, 46, 99, 121, 126
  see also individual species
Cluster clover 57

Cobalt 30, 78, 134
Cocksfoot 22–4, 41, 88
Continuous grazing 148
Cool temperate zone 8, 9, 10
Copper 78, 134
Costs 11, 15, 16–17, 46, 56, 77, 171, 174, 179
Couch grass 126
Cover crops 77
Crude protein 8, 9, 160, 161–2
Cultural control, weeds 115

D
Dairy pastures 32
   feeding 162
   tainting 152
Deferred grazing 148
Desmodium 46
Digestibility 8, 13, 26, 37, 157–61
Direct drilling 16, 75, 145
Docks 17, 84, 116–17
Drought, effects 11, 14, 69, 77
   species 28, 39, 41, 66

E
Equipment, chemical 128–33
   hay 165–9
   pasture 11, 15–16, 75, 77, 78, 81–3, 114
   silage 174–5, 179
Ergot 26
Erosion 80, 120, 126, 136
   water 15
   wind 12, 15, 28

F
Fabaceae 45
Feeding, strategies 150, 157
Fertiliser, application 145
   management 13, 26, 51, 56, 78, 82, 134–46
Fertility, effects 11
Field cricket 94–5
Fodder crops 100, 150–7
Formaldehyde 158, 174
Formic acid 174
Formononetin 46, 55
Frost 12, 14
   tolerance 37
Fungal diseases 26, 32, 55, 57, 87, 117
Fungicides 32
Fusarium wilt 67

G
Germination 10, 12, 14, 18–19
  requirements 73–4
Grazfeed 149
Grazing, damage 11, 64, 76
  strategies 10, 14, 22, 30, 75, 82, 83, 116, 122, 146–9, 152
Growth patterns 7–9
Gypsum 13

H
Hard seed 18
Harrowing 76, 83
Hay, baling 166–9
  cutting 159–64
  digestibility 158–62
  pastures 20, 37, 45, 64, 76, 157, 160
  storage 166–9
Hay itch mite 161
Herbicides 26, 37, 75, 79, 82, 124–33
  action 84, 124–7
  non-selective 15
  selective 15, 116
Horehound 107, 108

I
Infertility 46, 70
Information sources 14, 17–18, 51–2, 87, 119, 133, 137, 144, 149
Innocent weed 113, 114
Inoculation 48–50, 82, 83, 157
Insecticides *see* Pesticides
Integrated Pest Control 85, 88
Irrigation, crops 22, 24, 26, 32, 37, 41, 64, 66, 67, 69, 70, 150, 155
  fertiliser 145
Italian ryegrass 37–9

K
Kabatiella *see* Clover scorch
Kangaroo grass 21
Kikuyu 10, 24–6

L
Lablab 46
Ladybird 91
Leaching 136
  of herbicides 125, 127
Leucaena 46
Ley farming 9–10
Lime 48–9, 78, 141
Lucerne 11, 13, 45, 46, 64, 66–8, 77, 123, 126
  as hay 160, 165
  pests 90, 93, 96, 101, 105
Lucerne flea 51, 56, 80, 96–7

M
Maize 150, 153–4
Mangels 151
Medics 34, 45, 46, 47, 55–7, 121, 126
  species 57, 58–9
  pests 91, 93, 96, 99
Mediterranean zone 7
Melelotus 57
Millet 150, 152–3, 154
Molasses 174
Mowing 15, 67, 75, 82, 164–5

N
Native budworm *see* Climbing cutworm
Native grasses 21
Nitrogen 45–6, 78, 134
Nitrogen fixing 9, 45–6, 48–50
Nodules *see* Rhizobium
Noogoora burr 113, 114
Nutrients 13
  deficiencies 137–9
Nutritive value 13, 37, 45

O
Oats 26, 64, 76, 77, 88, 150, 155–6, 160, 161
Organic carbon 137, 141
Organic matter 13

P
Palatability 13, 20, 22, 26, 30, 39, 64, 107, 116, 122
Paspalum 26–7, 155
Pasture cockchafers 97–9
Pasture mixtures 13, 16, 34, 51, 155
Paterson's curse *see* Salvation Jane
Pea aphid 56, 67, 99–100
Peas 93, 155
Perennial veldt 11, 20, 28, 29
Perennials 10
Persian clover *see* Shaftal
Pest control 80, 85–106
Pesticides 85, 88
  recommendations for use *see* individual pests
  safety 85–6
  timing 87 *see also* Spraying
Pests, damage 11, 15, 64, 77, 84
  see also individual species
pH 13, 46, 78, 137, 139–41
Phalaris 22, 28, 30–2, 41, 77, 107, 109, 152, 160
Phalaris, staggers 30
Phalaris, sudden death 30
Phosphatic fertilisers 56, 83, 142–3
Phosphorus 48, 78, 134, 142
Pink cutworm 100–1

Index

Ploughing 76
Potassium 78, 134, 141
Potato weed 107, 111
Prairie grass 32–3
Prickly pear 118
Pro-Graze 149
Puccinellia 34–5
Pugging 34, 80

R

Rainfall patterns 7–10, 12, 21
Raking 83, 165
Rape 150, 151–2
Red clover 37, 70, 72, 160
Red-legged earthmite 51, 56, 80, 97, 101–3
Renovating 10, 12, 146–7, 150
Residual effects 119
Resistance, disease 14, 67
    pest 14, 56, 67, 85, 88, 91, 97, 100, 103, 104, 106
Rhizobium 45, 48–9, 104
Rolling 76–7
Root rot 67
Ropewick 123–4, 129
Rose clover 58–9, 61
Rotational grazing 66, 148
Ryegrass, annual 34, 35–7, 64, 88, 107, 109, 156, 159, 160, *see also* Italian ryegrass
    perennial 22, 26, 30, 39–41
Ryegrass staggers 39

S

Safety 85, 119, 121, 130–3, 178
    clothing 86, 174
Salinity 13, 143
Salt tolerance, species 13, 34. 41, 43, 57, 66, 151
Salvation Jane 84, 107, 111, 116, 122, 126
Sarradella 59, 61–2
Seed, costs 16
    quality 17–19
Seed set 10
Seedbed preparation 74
Selecting species 10–15, 46, 52, 55, 67
Set stocking 148
Shaftal clover 37, 39, 61, 63–4
Silage 45, 160, 170–9
    cutting 171–2
    making 170–5
    storing 175–8
Silver grass 120
Silverleaf nightshade 127
Sirato 46
Site 11–12, 157–8
Sitona weevil 56, 103–4

Skeleton weed 118, 127
Smut 32
Snout mite 97
Sod seeding 13, 15, 16, 34, 39, 51, 79–82
Soil, aeration 74, 77–8, 137
    fertility 9, 13, 48
    nutrients 74, 134–8 *see also* Soil fertility
    microbiology 13, 45
    structure 13, 16, 45, 74
    testing 139–45
    type 12–13
Sorghum 150, 153, 154–5
Sorrel 84
Soursob 107, 108, 126
Southern armyworm *see* Barley grub
Sowing 39, 56, 78–84
    timing 14–15, 69, 73–4
Spotted alfalfa aphid 56, 67, 99, 105–6
Spray grazing 15, 84, 122
Spray topping 15, 84
St John's wort 118
Stocking rates 9, 148
Storksbill 107, 110, 116
Strawberry clover 46, 48, 69-70, 71, 94
Strip grazing 148
Stylo 46
Subterranean clover 34, 46, 47, 51–5, 56, 145, 160
    identification 51–2
    pests 93, 96, 99
Sudax 153
Superphosphate *see* Phosphatic fertilisers
Surfactants *see* Wetting agents

T

T boot 81
Tagasaste 46
Tall fescue 41–3, 88
Tall wheatgrass 43–4
Temperature, soil 12
Thistles 116
Three cornered jack 112, 114
Time control grazing 148
Tissue analysis 144–5
Top dressing 145, 146
Topography 7, 11
Toxicity 14, 22, 26, 30, 37, 107, 116, 122, 153, 154
Trace elements 134–5, 143
Triticale 156
Tropical species 21
Tropical zone 8, 9, 46
Turnips 150, 151–2

U

Urea 145

V

Varietal purity 17–18
Varieties 13
Vetch 64, 65, 155, 156–7
Vigour 28, 107, 136
   seedling 10, 14, 15, 24, 82

W

Wallaby grass 21
Wallace plough 77
Warm temperate zone 7, 9, 20, 48
Waterlogging 8, 13, 14, 28, 43, 48, 57, 64, 69
Weeds 10, 12, 15, 20, 37, 57, 69, 114
   control 74, 75–6, 79–80, 82, 83, 107, 133, 146, 150 *see also* Chemical ploughing, individual weeds
   effects 77, 107, 114
   noxious 17, 114, 119
Weeping grass 21
Wetting agents 122, 127–8
Wheat 76, 88, 156, 160, 161
White clover 46, 67, 69, 70, 71
Wool contamination 20, 56

Y

Yeomans plough 77

Z

Zinc 78, 134